Studies in Logic
Volume 91

A View of Connexive Logics

Volume 81
Factual and Plausible Reasoning
David Billington

Volume 82
Formal Logic: Classical Problems and Proofs
Luis M. Augusto

Volume 83
Reasoning: Games, Cognition, Logic
Mariusz Urbański, Tomasz Skura and Paweł Łupkowski, eds.

Volume 84
Witness Theory. Notes on λ-calculus and Logic
Adrian Rezuş

Volume 85
Reason to Dissent. Proceedings of the 3rd European Conference on Argumentation. Volume I
Catarina Dutilh Novaes, Henrike Jansen, Jan Albert van Laar and Bart Verheij, eds.

Volume 86
Reason to Dissent. Proceedings of the 3rd European Conference on Argumentation. Volume II
Catarina Dutilh Novaes, Henrike Jansen, Jan Albert van Laar and Bart Verheij, eds.

Volume 87
Reason to Dissent. Proceedings of the 3rd European Conference on Argumentation. Volume III
Catarina Dutilh Novaes, Henrike Jansen, Jan Albert van Laar and Bart Verheij, eds.

Volume 88
Belief Attitudes, Fine-Grained Hyperintensionality and Type-Theoretic Logic
Jiří Raclavský

Volume 89
Essays on Set Theory
Akihiro Kanamori

Volume 90
Model Theory for Beginners. 15 Lectures
Roman Kossak

Volume 91
A View of Connexive Logics
Nissim Francez

Studies in Logic Series Editor
Dov Gabbay   dov.gabbay@kcl.ac.uk

A View of Connexive Logics

Nissim Francez

© Individual author and College Publications, 2021
All rights reserved.

ISBN 978-1-84890-370-8

College Publications
Scientific Director: Dov Gabbay
Managing Director: Jane Spurr

http://www.collegepublications.co.uk

All rights reserved. No part of this publication may be reproduced, stored in a retrieval system or transmitted in any form, or by any means, electronic, mechanical, photocopying, recording or otherwise without prior permission, in writing, from the publisher.

Contents

1 Introduction   **1**

    1.1 Motivation and background 1

        1.1.1 An interlude: motivating logics by natural language phenomena 2

        1.1.2 Connexivity 5

    1.2 The revolt against Classical logic 6

    1.3 Notational and terminological conventions 16

2 Characteristics of Connexive logics   **19**

    2.1 Introduction 19

    2.2 The propositional connexive axioms 19

    2.3 First-order Connexive logics 27

    2.4 Negating the conditional 28

    2.5 Rejecting classical validities 33

	2.5.1	Rejecting conjunction simplification	35
	2.5.2	Rejecting disjunction addition	36
	2.5.3	Rejecting contraposition	38
2.6	Objections to the connexive characteristics		39
2.7	Empirical corroboration of connexivity		41

3 A sample of connexive logics   43

3.1 How are connexive logics obtained? 44

3.2 Wansing's propositional Connexive logic C 45

     3.2.1 Axiomatic presentation of C 46

     3.2.2 Axiomatic derivability of the connexive axioms . 47

     3.2.3 Complete models for C 48

     3.2.4 Model-theoretic establishment of the connexive axioms 50

     3.2.5 Paraconsistency and paracompleteness of C . . . 55

     3.2.6 Classical validities invalidated by C 57

     3.2.7 Variations on C 57

3.3 Priest's propositional Connexive logics 60

     3.3.1 Models for P_S 61

     3.3.2 Connexivity of P_S 63

	3.3.3	Paraconsistency and paracompleteness of P_S	64
3.4	Angell's propositional Connexive logic		65
	3.4.1	Introduction	65
	3.4.2	The axiomatic presentation of $\mathbf{PA_1}$	65
	3.4.3	4-valued truth-tables for $\mathbf{PA_1}$	68
3.5	Francez's Connexive logics		72
	3.5.1	Introduction	72
	3.5.2	Natural language motivation	74
	3.5.3	Ramsey's test and a dual test	80
	3.5.4	Negating atomic propositions	81
	3.5.5	The natural-deduction systems $\mathcal{N}^{\neg r}$ and $\mathcal{N}^{\neg l}$	81
	3.5.6	The natural-deduction system $\mathcal{N}^{\neg l}$	88
	3.5.7	Model-theory for $\mathcal{N}^{\neg r}$ and $\mathcal{N}^{\neg l}$	91
	3.5.8	Conclusion	95
3.6	Some more Connexive logics		96
	3.6.1	Nelson	97
	3.6.2	McCall	97
	3.6.3	Pizzi and Williamson	97
	3.6.4	Estrada-González and Ramírez Cámara	98

	3.6.5	Wansing and Unterhuber	98
	3.6.6	Omori	98
	3.6.7	Rahman and Rückert	99

4 The scope of connexivity — 103

4.1 Introduction 103

4.2 Restricting the scope of connexivity: Humble connexivity 104

4.3 Extending the scope of connexivity: Poly-connexivity . 106

    4.3.1 Introduction 106

    4.3.2 Axiomatic definition of $PCON$ 108

    4.3.3 Models for $PCON$ 109

    4.3.4 The rules of $\mathcal{N}_{PCON}$ 111

    4.3.5 Connexivity of $PCON$ 114

    4.3.6 Deductive equivalence of $PCON$ and $\mathcal{N}_{PCON}$. 115

    4.3.7 NL motivation of $PCON$ 118

5 Connexivity and relevance — 129

5.1 Introduction 129

5.2 Connexive Relevant logic 130

    5.2.1 Introduction 130

		5.2.2	The natural-deduction system $\mathcal{N}_{rc}$	131
		5.2.3	On the negated conditional	140
		5.2.4	A natural deduction-theorem for $\mathcal{L}_{rc}$	141
	5.3	Axiomatic definition of $\mathcal{L}_{rc}$		143
		5.3.1	Defining $\mathcal{H}_{rc}$	143
		5.3.2	Deductive equivalence of $\mathcal{H}_{rc}$ and $\mathcal{N}_{rc}$	145
	5.4	Conclusion		147

6 An application of Connexive logics 149

	6.1	Introduction		149
	6.2	Connexive class theory		150
		6.2.1	Restricted quantification	150
	6.3	Defining classes in first-order Classical logic		152
	6.4	Connexivity and class inclusion		155
		6.4.1	Connexive algebra	155
	6.5	QC - Wansing's first-order connexive logic		158
		6.5.1	Defining classes in QC	161
	6.6	Priest's first-order connexive logics		167
		6.6.1	First-order QP_S	167
		6.6.2	Defining classes in QP_S	169

6.7	Another variant of Priest's connexive logic	171
6.8	Conclusions	174

7 Connexivity and modality — 177

7.1	Introduction	177
7.2	Weakening the connexive axioms	178
7.3	Defining the modal operators	179
7.4	Connexive modal logics	180
	7.4.1 Wansing's CK connexive modal logic	181
7.5	Francez' Connexive modal logic $CS5$	185
	7.5.1 Introduction	185
7.6	The connexive modal logic $CS5$	186
	7.6.1 The sequent calculus $SCS5$	187
	7.6.2 Derivability of the $S5$ axioms	187
	7.6.3 Derivability of connexive $S5$-non-theorems	189
	7.6.4 About $SC5$ models	190
	7.6.5 Linguistic support for $CS5$	190
	7.6.6 Possibility	194

8 Conclusion — 197

8.1	Uncovered topics		197
8.2	Topics for further research		198
	8.2.1	Connexivity and content	199
	8.2.2	Connexivity and bilateralism	200
	8.2.3	Provable contradictions	202

9 Glossary: linguistics terms   **205**

To Tikva, for a life-long connexive love

Preface

As an undergraduate student of mathematics and philosophy, when first having encountered the truth-functional connectives of Classical logic, I was immediately sceptic about them, especially about the material conditional. Years later, when I have encountered Relevance logics, and even more so when becoming acquainted with the axioms of Connexive logic, I felt confident that those approaches to logic are more adequate than the Classical logic. The adequacy criterion I have in mind is being a better approximation to inference and reasoning as carried out in natural language.

My main point in writing this book is to present a *snapshot*, certainly not the whole panorama, of a hot and emerging topic in the general area of *philosophical logic*, namely *Connexive logics*, hopefully having this topic gain popularity among a larger circle in the community. This is an important task especially in view of the absence of *any* book on this topic. The only source for Connexive logics currently available is Wansing's excellent encyclopedic entry [122], which was both inspiring and helpful in producing this book.

While presenting Connexive logics, I have, on passing, touched upon several other issues related to non-classical logics, to which Connexive logics belong.

Naturally, the presentation in the book is from a highly subjective point of view, putting emphasis on what *I* find interesting and intriguing in Connexive logic, without any ambition of a *full* coverage of the material available in the literature. The subjectivity is reflected by the more detailed coverage of my own work on Connexive logics, emphasizing the relationship with natural languages' informal logic. Some of the issues not covered, accompanied by reference to resources that do cover them, are listed in the Conclusion section.

In addition to the emphasis on NL, my point of view occasionally diverges from the orthodoxy on Connexive logics. As a notable example, I present in detail my more radical view of the scope of connexivity, extending it to more connectives than the traditional restriction of connexivity to the conditional in relation to negation.

Acknowledgments

My sincere thanks go, first of all, to Heinrich Wansing, the correspondence with whom over the years, both on topics related to Connexive logics and others, was enlightening and helpful.

I also thank to the following colleagues were also helpful in various ways: Thomas Ferguson, Paul Egré, Luis Estrada-González, Hitoshi Omori, Andreas Kapsner, Wolfgang Lenzen, Claudio Pizzi and Yale Weiss.

Many thanks to Jane Spurr of College Publications for devoted technical assistance.

Nissim Francez
Computer Science Dept., Technion, Haifa 32000, Israel.
francez@cs.technion.ac.il

Chapter 1

Introduction

1.1 Motivation and background

Connexive logics have their roots in antiquity, starting from ancient Greek philosophy, through middle age philosophy and enlightenment period philosophy, up to current philosophy, in particular *Philosophical logics*. I am *not* going to deal with any historical aspects of Connexive logics. The interested reader might find detailed historical descriptions in Wansing [122], McCall [74], Martin [70], or Lenzen [60].

A central[1] motivation in the development of Connexive logics, the one by which I myself am driven, is to formalize properties of the natural language (NL) conditional[2], properties not satisfied by the material conditional, the conditional of Classical logic (more on this in Section 1.2).

---

[1] Though not the only one. See Wansing [122] for other motivations.
[2] The NL conditional comes in (at least) two brands: *indicative* and *counterfactual*. I focus here on the former, with only occasional comments about the latter. For some discussion of counterfactuals and connexivity see Kapsner [57].

A disclaimer: My understanding of 'connexivity' is broader than than the standard use of this term in the literature, a broadening resulting from my emphasis on the relationship of connexivity and natural language. The common use of this term refers to a certain interaction of the conditional and the negation. I perceive connexivity as pertaining to other connectives too. This issue appears in several places throughout the book, for example in Chapter 4, mainly in Section 4.3.1, where justifications of my conceiving of connexivity is presented in more detail.
(end of disclaimer)

1.1.1 An interlude: motivating logics by natural language phenomena

There is a methodological question of how to use different linguistics resources as means for motivating and justifying logics. This question is hardly ever discussed explicitly in the literature. Below, I briefly present my own view on this methodological question.

Natural languages have, depending on specific theories of them, several layers in which various meaning related phenomena exhibit themselves. Besides the obvious layer of *semantics*, there are also the layers such as *pragmatics* and *felicity conditions*. In contrast, formal languages underlying logics have, in addition to syntax, just the single additional layer of semantics.

Pragmatics: At the pragmatic layer, there are phenomena such as *presupposition* and *implicature*.

> **presupposition:** A sentence like
>
> > (1) John ceased smoking

1.1 Motivation and background

presupposes

(2) John used to smoke

Here (2) has to be true in order for (1) to be either true or false.

Presupposition is considered as a motivation for some three-valued logics.

implicature: An implicature of

(3) some girl smiled

is that *not all* girls smiled.

More on the impact of pragmatics - see Section 2.5.2.

Felicity conditions: A felicity condition, for example, is the well-known Hurfords constraint [47], according to which in a felicitous disjunction no disjunct should entail the other. For example,

(4) John lives in Paris or in France

is infelicitous, in spite of its semantic acceptance.

Another important difference between natural and formal languages is that the former, serving as means of *communication*, assume the presence of a speaker and a hearer; in particular, the latter may be a participant in a *dialogue*. The presence of a speaker gives rise to aspects of meaning exhibited in speech only, in particular, *intonational stress*.[3]

None of this is present in formal languages.

As I see it, logics are *abstractions* of natural languages. As such, *all* those meaning related aspects of natural languages are eligible for being reflected in logical formal languages, both at the level of proof-system and at the level of model-theory, no other levels being available. Note

---
[3] As far as I know, intonational stress has not been used before to motivate logics.

that such a reflection can exhibit itself as additions to logics, modifying logics and rejecting from logics. For the latter, see Section 2.5. The former two occupy most of the book.

I use the meaning of the conditional in natural language, in particular the way it is negated in a dialogue, to motivate the axioms of Connexive logics. Furthermore, I use certain intonational stress variations to motivate a connexive theory of my own for conjunction, disjunction and modalities.

The following quotation from Cooper [14, p. 318] expresses an attitude towards motivating logic by NL usage, an attitude I share and endorse wholeheartedly:

> Many mathematical logicians adopt a distrustful stance toward any involvement of logic with natural language; they point out that one of the motivations for inventing formal languages in the first place was to escape from the idiosyncrasies of the natural languages. Although there may be some historical justification for it, the writer cannot help feeling that this attitude has outlived its usefulness. Near the turn of the century the question of interest was, "Does a precise, rational, and adequate logical system exist?" Now the question is more, "Which of the many possible, precise, rational, and adequate logical systems is best?" Surely there is something to be learned on this score from a logical system which evolved naturally through centuries of actual use by an advanced civilization. Even if there were not, it must still be admitted that, for, better or worse, man's thought processes are inextricably bound up with his native language. The question is therefore: Are we to choose a formal system which is as theoretically simple as possible, or rather one which is as consonant as possible with natural thought processes?

1.1.2 Connexivity

The following properties of a conditional (in relation to negation[4]), or of a logical consequence relation, underlie Connexive logics:

- **Con1: A proposition should not imply, nor be implied by, its negation.**

  In other words, accepting the formulas $\varphi \rightarrow \neg\varphi$ and $\neg\varphi \rightarrow \varphi$ seems "defective" in some sense, and this defect needs to be corrected in some way.

- **Con2: If a proposition φ implies some other proposition ψ, then the proposition φ should not imply also the negation of the proposition ψ, and vice versa.**

  Again, the joint acceptability of $\varphi \rightarrow \psi$ and $\varphi \rightarrow \neg\psi$ seems "defective", another defect that needs correction in some way.

Those principles, precisified below, are seen by many very plausible, and should be adhered to by any adequate theory[5] of the meaning of conditionals cast as a logic.

Remark 1.1.1 (terminology). *Throughout the book:*

- *I use 'proposition' as what is expressed by a formula in a logic, left further unspecified; in particular, it is* not *taken as a set of possible worlds. Mostly, the distinction between a formula and the proposition it expresses will be ignored by abuse of notation whenever no confusion will occur.*

---

[4]There is a vast literature about negation, too extensive to cite. The negation featuring in this book is very much geared to uses of negation in natural language. For a very recent exposition of negation in formal languages see Gabbay [38].

[5]in a recent paper [57], Kapsner argues that the source of the plausibility of the connexive principle resides in the *pragmatics* of the use of the conditional in NL, not from the meaning of the conditional.

- *I use 'conditional' and 'implication' (when used as a technical term) as synonyms.*

□

Example 1.1.1 (non-implication). *The following are NL-instances[6] exemplifying the above connexive principles.*

It is not the case that if it is raining then it is not raining

It is not the case that if it is not raining then it is raining

if it is the case that if it rains the match will be canceled, then it is not the case that if it rains the match will not be canceled

If production is rising, profits are not declining.
Therefore,
it is not true that if production is rising, profits are declining (i.e not rising).

□

1.2 The revolt against Classical logic

Almost from the moment of its inception, Classical logic was the dominant logic, very strongly entrenched among logicians and various "users" of logic in application areas such as mathematics, philosophy, linguistics, computer science and more. Consequently, its theses (formal theorems) were endowed with the title *tautologies*, and some of them even *logical truths*. When someone refers to 'logic' without further qualifications, it is typically Classical logic that is the referent.

However, in later periods a lot of dissatisfaction with Classical logic emerged, along a diversity of lines. As a result, a class of *non-classical*

---

[6]Abstracting away context dependence such as dependence on time and location.

1.2 The revolt against Classical logic 7

logics emerged, that includes a large and diversified number of logics, and *Connexive logics* amongst them.

As a result of the emergence of this variety of non-classical logics, a "hot" modern debate is between *logical monism*, endowing a single logic the status of "the (only) correct logic", in contrast to *logical pluralism*, denying the existence of a single "correct" logic". See Russell [107] for an overview.

Some of those non-classical logics were proposed as *rival logics*, intended to *replace* Classical logic as the "correct" logic. Other non-classical logics were seen as supplementary logics, dealing with issues not covered by Classical logic.

Remark 1.2.2 (extension vs. expansion). *A logic over some object-language is usually seen as a collection of formal theses (or validities); or, more generally, a collection of provable (or valid) sequents, involving also assumptions. For simplicity, I assume the former case.*

There are two important ways according to which logics in general, and Connexive logics in particular, are obtained[7] from other logics.

expansion: *A logic $\mathcal{L}_2$ expands another logic $\mathcal{L}_1$ when both logics share the same object-language, but $\mathcal{L}_2$ has additional formal theses.*

extension: *A logic $\mathcal{L}_2$ extends another logic $\mathcal{L}_1$ when the object-language of L_2 properly includes that of L_1 and $\mathcal{L}_2$ has more formal theses than $\mathcal{L}_1$; some of the additional theses involve operators from the larger object-language.*

$\mathcal{L}_2$ conservatively extends $\mathcal{L}_1$ if any formal thesis of $\mathcal{L}_2$ that involves only the object-language of $\mathcal{L}_1$ is also a formal thesis of $\mathcal{L}_1$.

There is a distinction between three main sorts of non-classical logics in terms of their relation to Classical logic, described below.

---
[7]This distinction is not always strictly adhered to in the literature.

Digression: Before describing this classification, I would like to comment on it. The point of departure of this classification, shared by many researchers in logic, is that Classical logic "is the measure of all things", and serves as the yardstick for evaluating any development. This is reflected even in the name of the family of logics in the classification to follow.

I, though, do not share this view. I believe the point of departure and the yardstick for measuring various logics is *how much do those logics capture (at least some) uses in natural language of the logics' connectives*.

In my view, Classical logic is just one of many logics attempting to capture some NL uses of the connectives, certainly not the most successful one for that task. □

The classification:

sub-classical logics: Those logics maintain the Classical logic vocabulary but *abandon* some of the logical validities of Classical logic.

supra-classical logics: Those logics extend the classical vocabulary[8] and *add* some logical validities.

contra-classical logics: Those logics also maintain the classical vocabulary, and both add some classical theses/validities *and* abandon others.

The classification in terms of logical validities is for convenience only. More generally, a similar classification is obtained in terms of logical consequence (involving assumptions).

---

[8]Because of the post-completeness of Classical logic, there is no way to only add logical validities without trivializing the logic.

1.2 The revolt against Classical logic

As will be seen in the sequel, Connexive logics are contra-classical.

Below is a short survey of qualms agains Classical logic, leading to several non-classical ones. This survey is not intended to be complete in any way, just to locate Connexive logics in a broader context.

Constructive logics: In Classical logic, one can prove an existential claim of the form $\exists x.\varphi(x)$, *without* pointing to any true instance $\varphi(a)$ of $\varphi(x)$. For example, one can use *reductio ad absurdum*, showing that the assumption $\forall x.\neg\varphi(x)$ leads to a contradiction. Similarly but more simply, one cannot establish a disjunction $\varphi \vee \psi$ *without* pointing out a true disjunct, a property called *the disjunction property*.

Applying such methods in mathematics, when reasoning about infinite sets, was rejected by the intuitionistic school of mathematics, the latter insisting on *constructions* of witnesses for existential claim.

Intuitionistic logic arose from formalizing those ideas, with its landmark rejection of the validity of the *excluded middle* law $\models \varphi \vee \neg\varphi$ and the *double negation* principle $\models \neg\neg\varphi \supset \varphi$, also in its rule form

$$\frac{\neg\neg\varphi}{\varphi} \ (DN)$$

According to the classification above, Intuitionistic logic is a subclassical logic.

Applying similar ideas to the *falsity* of $\forall x.\varphi(x)$, requiring the construction of a false instance $\varphi(a)$ of $\varphi(x)$, led Nelson [77] to the formulation of N4, a logic with *constructive negation*. Instead of the classical clause

$$\mathcal{M} \models \neg\varphi \text{ iff } \mathcal{M} \not\models \varphi$$

falsity is attributed to atomic propositions *independently* of their truth attribution, and then propagated over other connectives and quantifiers.

Similarly to the disjunction property, logics with strong negation have the dual property: one cannot falsify a conjunction without falsifying one of the conjuncts.

We will encounter the idea of strong negation in the devising of some Connexive logics in the sequel.

Paraconsistent and paracomplete logics: There is an important distinction between two notions of a *inconsistency* of a logic, say $\mathcal{L}$:

> **negation inconsistency:** This is the property of $\mathcal{L}$ validating, or deriving, a *contradiction*, a formula of the form $\varphi \wedge \neg \varphi$ for some φ. Or, in the absence of '$\wedge$' from the object-language, a set Γ such that for some φ $\{\varphi, \neg\varphi\} \subseteq \Gamma$.
>
> **Post inconsistency:** This is the property of $\mathcal{L}$ of validating, or deriving, *every* formula φ. A Post-inconsistent logic is also called *trivial*.

For Classical logics, both notions coincide, but they need not coincide in general.

According to the classification[9] of the meaning of a contradiction in Priest [96] (to which I return in more detail in Section 3.3), three main levels of logical consequence from a contradiction can be distinguished.

> **total:** From a contradiction *everything* follows (a property of logical consequence called *explosion*), also known as *Ex Contradictione Quodlibet (ECQ)*. This, for example, is the situation in Classical logic and Intuitionistic logic.
>
> **partial:** From a contradiction *not everything* follows. Typical logical consequences of a contradiction $\varphi \wedge \neg \varphi$, according to the partial account, are: φ, $\neg\varphi$ and $\varphi \wedge \neg \varphi$ itself. More generally, every $\varphi \in \Gamma$ is a logical consequence of an inconsistent Γ.

---
[9]The origin of this classification is Routley&Routley [106].

1.2 The revolt against Classical logic

null: From a contradiction *nothing* follows, not even itself, known as *Ex Contradictione Nihil (ECN)*. For a recent defense of this position, see Hewitt [45].

Analogous three levels can be distinguished as to what is a tautology (a logically valid proposition) a logical consequence of.

total: Every tautology is a logical consequence of *everything*, as in Classical logic (a property called *implosion*).

partial: A tautology is not a logical consequence of *everything*. For example, a tautology might be a logical consequence of itself.

null: A tautology is a logical consequence of *nothing*, not even of itself.

Recall that the definition of logical consequence in Classical logic is by *necessary truth preservation* from assumptions to conclusion. As a result, the following two objectionable properties of logical consequence hold:

- For *every* φ, ψ:
$$\varphi \wedge \neg \varphi \models \psi \tag{1.2.1}$$
  or, without conjunction
  $$\varphi, \neg \varphi \models \psi$$
  That is, *every* proposition is a logical consequence of a contradiction (or a contradictory set of assumptions), the total account mentioned above, leading to explosion.

- Analogously,
$$\varphi \models \psi \vee \neg \psi \tag{1.2.2}$$
  That is, *any tautology* (exemplified[10] by $\varphi \vee \neg \varphi$ above) is a logical consequence of *any* proposition, the total account again, leading to implosion.

---
[10]For expressing the property without using disjunction, a multiple conclusions formulation is required.

The objection to the above properties of logical consequence is that for an appropriate notion of logical consequence, *some connection in content* between the assumptions and the conclusion need to be present. Below (for Relevance logics), similar objections are raised in the context of the conditional connective, the internalization of logical consequence into the object-language.

I regard the reduction, as far as logic is concerned, of the contents of a formula merely to its truth-value, as done in Classical logic, as an over-generalization of the meanings of natural language affirmative sentences.

Those objections gave rise to logics which are *negation inconsistent but not trivial*: *not* validating the above two properties of logical consequence, called *paraconsistent – evading (1.2.1)*, and *paracomplete – evading 1.2.2)*. Typically, Connexive logics are both paraconsistent and paracomplete, containing contradictions but being non-trivial. See, for examples, propositions 3.2.6 and 3.2.7 below. For an overview of paraconsistency, see Priest [97] and Béziau, Carnielli and Gabbay [11].

One should carefully distinguish between paraconsistency, the evading of explosion, and *dialetheism* – the philosophical position maintaining that *some contradictions are true*. Or, under some other approaches, are *both true and false*. Devising a logic validating some contradictions is a way to obtain a paraconsistent logic, but such a logic is not necessarily tied to dialetheism.

Relevance logics: Two of the most objectionable theses of Classical logic pertain to the material conditional, expressing the "facts" that:

1. A true proposition is implied by *any* proposition.

$$\text{For every } \varphi, \psi: \ \vdash_{cl} \varphi \supset (\psi \supset \varphi) \quad (1.2.3)$$

2. A false proposition implies *any* proposition.

$$\text{For every } \varphi, \psi: \ \vdash_{cl} \neg \varphi \supset (\varphi \supset \psi) \quad (1.2.4)$$

1.2 The revolt against Classical logic

Jointly, those theses were coined as "the paradoxes of the material implication".

Another objectionable property of the material conditional is

$$\text{For every } \varphi, \psi: \quad \vdash (\varphi \supset \psi) \vee (\psi \supset \varphi) \qquad (1.2.5)$$

That is, for *any* two propositions, one of them implies the other. A particularly irritating instance of (1.2.5) is

$$\text{For every } \varphi: \quad \vdash (\varphi \supset \neg\varphi) \vee (\neg\varphi \supset \varphi) \qquad (1.2.6)$$

Relevance logics, originating from Anderson and Belnap [1], revolt against the truth-functionality of the conditional, maintaining that in a valid conditional, *some connection* in contents is required between the antecedent and the consequent. The least requirement of such connection in contact, still relating to truth, is a *dependence* of the truth of the consequent on the truth of the antecedent. The mere truth of the former on the truth of the latter cannot establish implication.

Such a requirement of connection in content invalidates the paradoxes of the material conditional as well as (1.2.5), amongst other effects it has.
In the classification above, Relevance logics are in general subclassical logics.

There are in the literature are two main embodiments of the relevance requirements.

1. In a valid conditional, the antecedent and the consequent *share* at least one atomic formula[11]. This is known as the *variable-sharing condition (VSC)*, regarding contents in a very syntactic way.

---
[11] Under this conception, atomic formulas are viewed as propositional variables.

2. In a proof of a conditional (possibly from some auxiliary assumptions) using a conditional proof[12], the assumed antecedent *must be used* (in some appropriate definition of usage) in deriving the consequent.
   In systems with structural rules, the rule of *Weakening* (adding an assumption) is inadmissible, as the additional assumption might be irrelevant to the conclusion.

A conditional satisfying the above properties is a *relevant conditional*.

The relevance requirement is also extended from the valid conditional to the logical consequence, where in $\Gamma \vDash \varphi$ the assumptions Γ have to be relevant to the conclusion φ.

A typical rule rejected by Relevance logics is *Ex falso quodlibet (EFQ)*

$$\frac{\bot}{\varphi} \ (EFQ)$$

allowing the inference of *any* φ from a contradiction.

I return to Relevance logic and their connection to Connexive logics in Chapter 5.

Modal logics: Modal logics are supra-classical logics, extending the object-language with two (possibly indexed) connectives, called modalities, expressing properties of propositions that transcend "plain" truth or falsity:

- $\Box \varphi$: read as φ is *necessary*.
- $\Diamond \varphi$: read as φ is *possible*.

There are many interpretations of the modalities; usually, the modalities are assumed to be *mutually dual*.

$$\Diamond \varphi \equiv \neg \Box \neg \varphi \qquad \Box \varphi \equiv \neg \Diamond \neg \varphi$$

---

[12]That is, Implication introduction in natural deduction systems.

1.2 The revolt against Classical logic

The most popular model-theory for modal logics is by means of Kripke models, based on *frames* that contain points[13] of evaluation, possibly connected via a binary *accessibility relation*.

- $\Box\varphi$ holds at a point of evaluation w iff φ holds at *every* point of evaluation w' accessible from w.

- $\Diamond\varphi$ holds at a point of evaluation w iff φ holds at *some* point of evaluation w' accessible from w.

Different modal logics, satisfying different axioms, can be obtained by imposing restrictions on the accessibility relation.

Modality is related to connexivity in Chapter 7.

Multi-valued logics: In contrast to the bivalence of Classical logic, using only the two truth-values t and f, multi-valued logics (known also as many-valued logics) employ a larger (possibly infinite) set $\mathcal{V}$ of truth-values. There is a variety of interpretations of the "extra" truth-values. For a review, see Gottwald [41].

Some Connexive logics are multi-valued; for example, Angell's P_{A1} (see Section 3.4.3) is defined over four truth-values. Other Connexive logics are extensions of multi-valued logics; for example, the Connexive Modal logic BK^- of Odintsov, Skurt and Wansing (see Section 7.4.1.1) extends the four-valued logic FDE (see Section 7.4.1.2). The latter has two extra truth-values b and n, where b (read 'both') represents being *both* true and false – a truth-value *glut*, while n (read 'none') represents being *neither* true nor false – a truth-value *gap*.

---

[13] A popular name for those points is "possible worlds", which I consider a misnomer.

1.3 Notational and terminological conventions

Initially, I consider a propositional core object-language generated by a countable set **At** of *atomic propositions*, ranged over by metavariables p, q, and closed under two connectives: '$\to$' - the conditional (or implication[14]) and '$\neg$' - the negation[15]. Occasionally, additional connectives are included. Metavariables φ, ψ range over arbitrary propositions[16].

By a systematic abuse of nomenclature, I use 'conditional' both for the connective '$\to$' itself and for formulas of the form '$\varphi \to \psi$', and similarly for 'conjunction' and 'disjunction' when present. No confusion should arise.

For a formula φ, $At[\![\varphi]\!]$ denotes the collection of all atomic propositions occurring in φ. Metavariables Γ, Δ range over finite sets of propositions.

The object-language of a first-order Connexive logic includes (possibly indexed) individual variables x, x_i, relation symbols (of implicit arity) P, R and the usual quantifiers $\forall, \exists$. Let $\overline{x}$ abbreviate $x_1, \cdots, x_n$ and $\forall \overline{x}$ – $\forall x_1 \cdots \forall x_n$, for some $n \geq 1$. The notions of free and bound variable, as well as open and closed formulas, are defined as usual.

I use '$\vdash$', possibly indexed by a name of a proof system, for *derivability* in the named proof system. Occasionally, if say, 'A' is a name for some formula, say φ, mostly an axiom, I use '$\vdash A$' by an abuse of notatIon to stand for '$\vdash \varphi$'.

The mention vs. use distinction I employ is the following. Mentioned

---

[14] In the sequel, I will use 'conditional' even where 'implication' is used in the literature.

[15] Occasionally, when reporting some work denoting negation by '$\sim$', I adopt the original notation.

[16] For simplicity, I ignore throughout the distinction between formulas (in the object-language) and the propositions those formulas express.

1.3 Notational and terminological conventions

words are always within single quotes, e.g., 'connexive'. When a single symbol is mentioned, it is also within single quotes, e.g., '→'. When a whole formula is mentioned, it is left unquoted.

Natural language expressions in examples iare always displayed in sanserif font, like Every man is mortal, and are always mentioned, not used.

Names of logics, such as 'Classical logic' or 'Connexive logic' are considered proper names and, hence, capitalized.

For quantified formulas I us what is known as the "dot notation", saving some parentheses. The scope of the preceding quantifier extends as far as possible, i.e., to the end of the formula unless it is blocked by an actual parenthesis.

Chapter 2

Characteristics of Connexive logics

2.1 Introduction

There is no *precise*, definite demarcation of Connexive logics. Thus, I speak only of their most common characteristics, considered as axioms[1] generally agreed upon in the community working on those logics. See Estrada-González and Ramirez Cámara [26] for a discussion of additional possible desiderata for Connexive logics.

2.2 The propositional connexive axioms

The connexive principles come in two guises:

---
[1]Those are actually axiom schemes, but I will refer to them throughout as axioms, for brevity.

- As features of the conditional connectives (the more popular guise).

- As features of the *logical consequence relation*.

I focus here on the first guise, turning to the second only occasionally.

In this section, the sign '⊢' denotes derivability in some *generic* Connexive logic.

The first two characteristics of propositional Connexive Logics, capturing the principle **Con1** above, are the following (formal) theorems (theses), which are *not* theorems of Classical logic.

$$\text{For every } \varphi: \quad \begin{array}{l} A_1: \ \vdash \neg(\varphi \to \neg\varphi) \\ A_2: \ \vdash \neg(\neg\varphi \to \varphi) \end{array} \quad (2.2.1)$$

Both are jointly known as *Aristotle's thesis*. In terms of the "defect" of $\varphi \to \neg\varphi$ and $\neg\varphi \to \varphi$ mentioned in **Con1**, the correction is to negate the defective formulas.

One way to paraphrase the contents of Aristotle's connexive axioms is: *the antecedent of a valid conditional must be compatible with the consequent*, and no proposition is compatible with its negation.

Note that in the presence of disjunction in the object-language, the A_i axioms invalidate the "dubious" disjunction in (1.2.6), which is a disjunction of the negated A_i axioms.

In a sense, those characteristic axioms differ from axioms in many other logics, because they can be interpreted as what *cannot* be proved. Indeed, as noted by Wansing [122], the same intuition can be captured, though less intuitively, by non-provability claims:

$$\text{For every } \varphi: \quad \begin{array}{l} A_1^\vdash: \ \nvdash \varphi \to \neg\varphi \\ A_2^\vdash: \ \nvdash \neg\varphi \to \varphi \end{array} \quad (2.2.2)$$

2.2 The propositional connexive axioms

Namely, the correction of the defect mentioned in **Con1** is to render the defective formulas unprovable.

As explained by Ferguson [29] (p. 26), the two formulations of the connexive intuition, the two ways of correcting the defect of $\varphi \to \neg\varphi$ and $\neg\varphi \to \varphi$ mentioned in **Con1**, have a subtle distinction. While the two occurrences of the object-language negation '$\neg$' in each axiom suggest they are *the same* negation, the meta-language negation in '$\nvdash$' may be a different negation than the object-language negation '$\neg$'. It seems this potential difference has not been explored in the literature.

Yet another perspective for viewing A_1 can be gained by the use of $\varphi \to \neg\varphi$ as a *formal contradiction* for the sake of a kind of reductio ad absurdum by McCall [71]. His Connexive logic $CC1$ (see Section 3.6.2) has a formal thesis

$$\text{for every } \varphi \text{ and } \psi: \quad \vdash_{CC1} (\varphi \to (\psi \to \neg\psi)) \to \neg\varphi$$

That is, if φ implies $\psi \to \neg\psi$, a "contradiction", then $\neg\varphi$ is implied.

The contra-classicality induced by the connexive axioms:

- A_1 is not a Classical logic theorem. Any classical model (a classical valuation) $\mathcal{M}$ s.t. $\mathcal{M} \models \neg\varphi$ also verifies $\mathcal{M} \models \varphi \supset \neg\varphi$, falsifying A_1.

- A_2 is not a classical logic theorem either. Any classical model $\mathcal{M}$ s.t. $\mathcal{M} \models \varphi$ also verifies $\mathcal{M} \models \neg\varphi \supset \varphi$, falsifying A_2.

The other connexive characteristic relationships, also non-theorems in Classical logic, are:

$$\text{For every } \varphi \text{ and } \psi: \quad \begin{matrix} B_1: & \vdash (\varphi \to \psi) \to \neg(\varphi \to \neg\psi) \\ B_2: & \vdash (\varphi \to \neg\psi) \to \neg(\varphi \to \psi) \end{matrix} \quad (2.2.3)$$

and are attributed in Kneale&Kneale [59] to the ancient Philosopher and logician Boethius. Those axioms capture the principle **Con2** above, correcting the defect mentioned in that principle, namely the joint acceptability of $\varphi \to \psi$ and $\varphi \to \neg\psi$ by having the former imply the negation of the latter.

Of a connexive flavor, also related to the principle **Con2**, are also the following axioms,

$$\text{For every } \varphi \text{ and } \psi : \quad \begin{array}{l} B_3 : (\varphi \to \psi) \to \neg(\neg\varphi \to \psi) \\ B_4 : (\neg\varphi \to \psi) \to \neg(\varphi \to \psi) \end{array} \quad (2.2.4)$$

as well as the *converses* B'_i of B_i, $i = 1, \cdots, 4$.

Below, we shall encounter a strengthening of all four the B_i axioms into equivalences.
First,

$$\text{For every } \varphi \text{ and } \psi : \quad \begin{array}{l} B_1^{\leftrightarrow} : \vdash (\varphi \to \psi) \leftrightarrow \neg(\varphi \to \neg\psi) \\ B_2^{\leftrightarrow} : \vdash (\varphi \to \neg\psi) \leftrightarrow \neg(\varphi \to \psi) \end{array} \quad (2.2.5)$$

Logics satisfying (2.2.5) were called by Sylvan (formerly Routley) [114] *hyper-connexive*.
Then, also

$$\text{For every } \varphi \text{ and } \psi : \quad \begin{array}{l} B_3^{\leftrightarrow} : (\varphi \to \psi) \leftrightarrow \neg(\neg\varphi \to \psi) \\ B_4^{\leftrightarrow} : (\neg\varphi \to \psi) \leftrightarrow \neg(\varphi \to \psi) \end{array} \quad (2.2.6)$$

There is also another formulation of the B_is, in the weaker rule form:

$$\text{For every } \varphi \text{ and } \psi : \quad \begin{array}{l} B_1^{\vdash} : \varphi \to \psi \vdash \neg(\varphi \to \neg\psi) \\ B_2^{\vdash} : \varphi \to \neg\psi \vdash \neg(\varphi \to \psi) \end{array} \quad (2.2.7)$$

See Wansing and Unterhuber [124], where such a logic is called *weakly connexive*.

The **Con2** principle can also be expressed as non-derivability requirements.

$$B_1^{\not\vdash} : \text{ for every } \varphi, \psi : \text{ if } \varphi \vdash \psi \text{ then } \varphi \not\vdash \neg\psi \quad (2.2.8)$$

2.2 The propositional connexive axioms

$B_2^{\not\vdash}$: for every φ, ψ: if $\varphi \vdash \neg\psi$ then $\varphi \not\vdash \psi$ \hfill (2.2.9)

and similarly for the converses of the B_is. This is another way of correcting the defect mentioned in **Con2**.

One can view also B_1 from the perspective of the kind of reductio ad absurdum mentioned above for A_1. A similar role of a formal contradiction for that kind reductio ad absurdum is played by $(\varphi \to \psi) \wedge (\varphi \to \neg\psi)$, in McCall [71] $CC1$, which has an formal thesis

$$\vdash_{CC1} (\varphi \to ((\psi \to \chi) \wedge (\psi \to \neg\chi))) \to \neg\varphi$$

That is, if φ implies $(\psi \to \chi) \wedge (\psi \to \neg\chi)$, a "contradiction", then $\neg\varphi$ is implied.

Digression: There are also other weak variants of the B_i axioms in an object-language having two conditionals, both the material '⊃' and the connexive '→':

For every φ and ψ:
$$\begin{array}{l} B_1^w : \vdash (\varphi \to \psi) \supset \neg(\varphi \to \neg\psi) \\ B_2^w : \vdash (\varphi \to \neg\psi) \supset \neg(\varphi \to \psi) \end{array}$$
\hfill (2.2.10)

Those axioms ere introduced by Pizzi [91] and Pizzi and Williamson [93]. See also Kapsner and Omori [58]. The motivation is capturing properties of *counterfactual conditionals*. a conditional which I will not consider in detail here any further. Those axioms were further studied by Weiss [125], in the context of extending (regular) Conditional logics. □

In case the object-language contains also (boolean) conjunction ('∧'), the following variant of the Boethius axiom is considered, attributed to Abelard.

For every φ and ψ:
$$\begin{array}{l} AB_1 : \vdash \neg((\varphi \to \psi) \wedge (\varphi \to \neg\psi)) \\ AB_2 : \vdash \neg((\varphi \to \psi) \wedge (\neg\varphi \to \psi)) \end{array}$$
\hfill (2.2.11)

When presented as properties of the logical consequence relation, Aristotle's axiom can be expressed as

For every φ: $\quad A_1^c : \varphi \not\vdash \neg\varphi \quad\quad A_2^c : \neg\varphi \not\vdash \varphi$

Similarly, Boethius' axioms can be presented as

$$B_1^c: \text{ if } \varphi \vDash \psi \text{ then } \varphi \nvDash \neg\psi \qquad B_2^c: \text{ if } \varphi \vDash \neg\psi \text{ then } \varphi \nvDash \psi$$

In addition to the A_i and B_i axioms, a negative characteristic is added, in the form of a non-derivability, to the effect that

$$\text{For every } \varphi \text{ and } \psi: \quad (asym) \nvdash (\varphi \to \psi) \to (\psi \to \varphi) \qquad (2.2.12)$$

The condition $(asym)$ is known as the asymmetry of the conditional, and is imposed to prevent interpreting the conditional as a biconditional.

The connexive axioms are valid under a truth-table with classical negation and a "conditional" defined as:

$$v[\![\varphi \to \psi]\!] = t \text{ iff } v[\![\varphi]\!] = v[\![\psi]\!]$$

See also Remark 2.4.6 below.

Some provision has to be assumed also w.r.t. negation, to avoid interpretations of negation such as $v[\![\neg\varphi]\!] = t$ independently of $v[\![\varphi]\!]$. Mostly, $\neg\neg\varphi \equiv \varphi$ is assumed.

Note that all the connexive principle pertain to *arbitrary* object-language formulas φ and ψ.
I return to this generality in Section 4.2.

Remark 2.2.3. *An immediate consequence of the validity of A_1 is that the principle known as 'ex impossibily quodlibet' (everything follows from an impossible proposition), namely*

$$\text{for every } \varphi \text{ and } \psi: \vdash \varphi \wedge \neg\varphi \to \psi \qquad (2.2.13)$$

cannot hold, because taking ψ as $\neg(\varphi \wedge \neg\varphi)$ yields

$$\vdash \varphi \wedge \neg\varphi \to \neg(\varphi \wedge \neg\varphi)$$

contradicting A_1.

2.2 The propositional connexive axioms

Similarly, A_2 forces the invalidation of another principle, 'necessarium ex quodlibet' (a necessary proposition follows from everything), for example

$$\text{for every } \varphi, \psi : \vdash \varphi \to \psi \vee \neg \psi \tag{2.2.14}$$

for taking φ as $\neg(\psi \vee \neg \psi)$ yields

$$\vdash \neg(\psi \vee \neg \psi) \to (\psi \vee \neg \psi)$$

contradicting A_2.

□

2.2.0.1 Interrelatedness of the connexive axioms

The connexive principles are not completely independent of each other.

Example 2.2.2. *Consider a logic in which $\vdash \varphi \to \varphi$ (identity) and containing the rule Modus Ponens*

$$\frac{\varphi \to \psi \quad \varphi}{\psi} \; (MP)$$

and also the uniform substitution closure.

We have

$$B_1 \vdash A_1$$

$$B_3 \vdash A_2$$

by substituting φ for ψ in B_i and detaching using identity.

□

Example 2.2.3. *Similarly, if a logic contains double negation equivalence and allows substitution of equivalents, than*

$$A_1 \vdash A_2 \text{ and } A_2 \vdash A_1$$

$A_1 \vdash A_2$ *is obtained by instantiating φ to $\neg\varphi$ in A_1, obtaining $\neg(\neg\varphi \to \neg\neg\varphi)$, and the using double negation equivalence. A similar reasoning yields $A_2 \vdash A_1$.*

Similarly, by instantiating ψ to $\neg\psi$, we get

$$B_1 \vdash B_2 \text{ and } B_2 \vdash B_1$$

□

2.2.0.2 Strong connexivity

In [54], Kapsner proposes a stronger version of the connexive principles, calling a logic respecting this strengthening as *strongly connexive*:

Unsat1: For every φ: both $\varphi \to \neg\varphi$ and $\neg\varphi \to \varphi$ are *unsatisfiable*.

Unsat2: For every φ and ψ: $\varphi \to \psi$ and $\varphi \to \neg\psi$ are not *simultaneously satisfiable*.

This is a stronger correction of the "defects" in **Con1** and **Con2**. In Estrada-González and Ramirez-Cámara [26] a Connexive logic satisfying **Unsat1** and **Unsat2** is called *Kapsner-strong* connexive logics.

As admitted by Kapsner, the Kapsner-strong connexivity maybe too strong of a property to require, and no "good" strongly connexive logics were presented by the time of this proposal.

2.3 First-order Connexive logics

As already mentioned above, Wansing [122] noted that the same intuition leading **Unsat1** and **Unsat2** leads also to (2.2.2), that can be considered a weaker versions of of this property (my own names), namely

WUnsat1: For every φ and ψ: $\nvdash \varphi \to \neg\varphi$ and $\nvdash \neg\varphi \to \varphi$.

WUnsat2: For every φ and ψ: $\varphi \to \psi \nvdash \varphi \to \neg\psi$ and $\varphi \to \neg\psi \nvdash \varphi \to \psi$.

2.3 First-order Connexive logics

First, a first-order Connexive logic will contain instances of A_i, B_i where φ is a first-order formula. For example, the following instance of A_1.

$$\vdash \neg(\forall x.\varphi(x) \to \neg(\forall x.\varphi(x))$$

More interestingly, I consider as the characteristics of first-order Connexive logics the formal theorems forming the natural lifting of the propositional characteristics in (2.2.1) and (2.2.3) to the first-order level. Typically, both $\varphi(\overline{x})$ and $\psi(\overline{x})$ are open formulas that have free occurrences of $\overline{x}$.

$$\begin{aligned} QA_1 &: \vdash \forall \overline{x}.\neg(\varphi(\overline{x}) \to \neg\varphi(\overline{x})) \\ QA_2 &: \vdash \forall \overline{x}.\neg(\neg\varphi(\overline{x}) \to \varphi(\overline{x})) \end{aligned} \quad (2.3.15)$$

$$\begin{aligned} QB_1 &: \vdash \forall \overline{x}.(\varphi(\overline{x}) \to \psi(\overline{x})) \to \neg(\varphi(\overline{x}) \to \neg\psi(\overline{x})) \\ QB_2 &: \vdash \forall \overline{x}.(\varphi(\overline{x}) \to \neg\psi(\overline{x})) \to \neg(\varphi(\overline{x}) \to \psi(\overline{x})) \end{aligned} \quad (2.3.16)$$

For example, suppose $P(x)$ expresses that x has the property denoted by P. Then, the following instance of A_1

$$\vdash \neg(P(x) \to \neg P(x))$$

expresses Aristotle's view that an object having P cannot imply that that object does not have P.

Remark 2.3.4. *As noted by Pizzi [91], QB_1 has an instance equivalent to*

$$\vdash \forall x.(\varphi(x) \to \psi(x)) \to \exists x.\varphi(x) \land \psi(x)$$

which, if '$\to$' is taken as the material conditional '$\supset$', reflects the Aristotelian implication between every A is B *and* some A is B, *an implication generally rejected by modern views, by Classical logic in particular.*

□

2.4 Negating the conditional

The following question is central to Connexive logics:

how should a conditional be negated?

The classical way of negating $\varphi \supset \psi$ by having $\neg(\varphi \supset \psi)$ be logically equivalent with $\varphi \land \neg\psi$ is found by many as inadequate, not reflecting the way the indicative conditional is negated in NL.

As will become evident as the presentation unfolds, an adequate way of negating the conditional *may lead* to the validity of the connexive axioms (see Proposition 2.4.1).

The inadequacy of the classical way of negating a conditional is persuasively argued for in the following dialogue from Cantwell [13], where he introduces a logic CN for *conditional negation*, though (not presented explicitly as a Connexive logic, though).

Example 2.4.4 (negating the conditional). *The following is from Cantwell [13] (p. 246):*

2.4 Negating the conditional

> ... *[C]onsider, for instance, the following exchange:*
> **Anne:** If Oswald didn't kill Kennedy, Jack Ruby did.
> **Bill:** No! You're wrong.
> *When Bill denies the conditional asserted by Anne, he neither asserts nor denies that Oswald did the killing (he can continue,* "If Oswald didn't kill Kennedy, Castro did"*); his denial seemingly amounts to no more than the assertion that if Oswald didn't shoot Kennedy then neither did Jack Ruby. This kind of "conditional denial" seems to be a basic move in the language game; conditional negation is the sentential operator that corresponds to this form of conditional denial:* "It is not the case that if Oswald didn't shoot Kennedy, Jack Ruby did." *So one reason for calling* '~' *"conditional" negation is that it seems to capture a particular way of negating conditionals.*

Cantwell's Example 2.4.4 is a good exemplification of this dissatisfaction with the classical way of negating the conditional. Thus, according to Cantwell's view as quoted above, the correct way to negate a conditional is according to the following equivalence:

$$\neg(\varphi \to \psi) \equiv \varphi \to \neg\psi \qquad (2.4.17)$$

the hyper-connexivity equivalence which we will encounter again and again in the sequel.

An informal interpretation of the equivalence (2.4.17) is a view of the conditional as *gappy*, lacking a truth-value in case the antecedent is false, where negation preserves gappyness.

The same idea, formulated somewhat differently (notation adjusted) is presented by Cooper [14] in his *ordinary language* logic OL:

$$\Gamma \vdash_{OL} \neg(\varphi \to \psi) \text{ iff } \Gamma, \varphi \vdash_{OL} \neg\psi \qquad (2.4.18)$$

The following dialogue is a variation on one presented by Cooper to exemplify negation the NL indicative conditional:

> A : If Jones retires before he is sixty,
> he will still retain all of his pension benefits.
>
> B : No! If Jones retires before he is sixty,
> he will not retain all of his pension benefits.

The source of this wrong capturing of the meaning of NL conditionals by Classical logic is that the relationship between the antecedent and the consequent of an NL conditional is depending on *content*, not merely on *truth* and *falsity*, the latter two being the only concerns of Classical logic. This observation extends from the conditional also to logical consequence.

In [2] (p. 336), Angell gives several pairs of NL sentences, instances of $\varphi \to \psi$ and $\varphi \to \neg \psi$, and refers to them as '*conflicting, logically incompatible sentences*'. For example,

> 2) If we had followed a different policy towards Germany in the 1920s, the second World War would not have occurred.
> 2') If we had followed a different policy towards Germany in the 1920s, the second World War would still have occurred.

When rendering 'if φ then ψ' as $\varphi \to \psi$, Strawson [113] (p. 452) sees the joint assertion of the conditionals $\varphi \to \psi$ and $\varphi \to \neg \psi$ (in the same context) as *self-contradictory*.

As yet another example, Downing [17] says[2]

> It seems clear, though it is perhaps impossible to prove, that subjunctive conditionals with the same antecedent and contradictory consequents cannot both be true.

---

[2] Quoted in Angell [2].

2.4 Negating the conditional

Also, in [18] (p. 491) Downing writes (about subjunctive conditionals):

> If one speaker utters the words 'If it rained, the match would be canceled' and another speaker replied with the words 'If it rained, the match would not be canceled', [...] two statements have been made that cannot, logically, both be true.

In addition to negating the conditional as in (2.4.17), we will encounter also another way of negating the conditional, harder to relate directly to negating NL conditionals.

$$\neg(\varphi \rightarrow \psi) \equiv (\neg\varphi \rightarrow \psi) \qquad (2.4.19)$$

This is one of the hyper-connexive axioms. Clearly, (2.4.17) is related to A_1 while (2.4.19) is related to A_2.

Remark 2.4.5. *In [53], Kamide and Wansing employ a negation scheme analogous to (2.4.19)*

$$\neg(\varphi \rightarrowtail \psi) \leftrightarrow (\neg\varphi \rightarrowtail \psi) \qquad (2.4.20)$$

for the co-implication connective '$\rightarrowtail$' (my notation) of the Bi-Intuitionistic logic BiInt in a connexive extension of Wansing's Connexive logic C (see Section 3.2 for the latter). □

Remark 2.4.6. *Pizzi and Williamson [93] observe that the satisfaction of (2.4.17) often tends to be accompanied by satisfaction of properties of a biconditional. Hence, imposing (2.2.12) to rule out interpreting the conditional as a biconditional is justified.* □

Another way to look at negating the conditional is through natural-deduction rules. In contrast to the *uniform way* of introducing and eliminating the negation $\neg\varphi$ in Classical logic by I/E-rules $(\neg I)$ and $(\neg E)$, independently of the negated formula φ, Connexive logic introduce and eliminate $\neg(\varphi\to\psi)$ by separate rules. There are two such sets of rules considered in Francez [33].

$$\dfrac{\begin{array}{c}[\varphi]_i\\ \vdots\\ \neg\psi\end{array}}{\neg(\varphi\to\psi)}\,(\neg\to I_r^i) \qquad \dfrac{\neg(\varphi\to\psi)\quad \varphi}{\neg\psi}\,(\neg\to E_r) \qquad (2.4.21)$$

and

$$\dfrac{\begin{array}{c}[\neg\varphi]_i\\ \vdots\\ \psi\end{array}}{\neg(\varphi\to\psi)}\,(\neg\to I_l^i) \qquad \dfrac{\neg(\varphi\to\psi)\quad \neg\varphi}{\neg\psi}\,(\neg\to E_l) \qquad (2.4.22)$$

The rules in (2.4.21) correspond to the axiom (2.4.17), while those in (2.4.22) correspond to the axiom (2.4.19). For more on those rules, see Section 3.5.

Similar sequent calculus rules for connexively negating the conditional can be found in Kamide and Wansing [51].

$$\dfrac{\Gamma:\Delta,\varphi \quad \Sigma,\neg\psi:\Pi}{\neg(\varphi\to\psi),\Gamma,\Sigma:\Delta,\Pi}\,(\neg\to L) \qquad \dfrac{\Gamma,\varphi:\neg\psi,\Delta}{\Gamma:\Delta,\neg(\varphi\to\psi)}\,(\neg\to R)$$

The following proposition establishes the fundamentality to Connexive logics of the way the conditional is negated.

Proposition 2.4.1 (negating the conditional and connexivity). *If a logic satisfies*

- (id) $\varphi\to\varphi$

- (dn) $\neg\neg\varphi\equiv\varphi$

- (rep) replacement of equivalents

then the connexive axioms A_i, B_i, $i = 1, 2$ are derivable from the negating the conditional equivalences (2.4.17) and (2.4.19).

Proof.

A_1:
$$\varphi \to \varphi \equiv \varphi \to \neg\neg\varphi \quad (dn)$$
$$\equiv \neg(\varphi \to \neg\varphi) \quad (2.4.17)$$

A_2:
$$\varphi \to \varphi \equiv \neg\neg\varphi \to \varphi \quad (dn)$$
$$\equiv \neg(\neg\varphi \to \neg\varphi) \quad (2.4.19)$$

B_1:
$$(\varphi \to \psi) \to (\varphi \to \psi) \equiv (\varphi \to \psi) \to \neg\neg(\varphi \to \varphi) \quad (dn)$$
$$\equiv (\varphi \to \psi) \to \neg(\varphi \to \neg\psi) \quad (2.4.17)$$

B_2:
$$(\varphi \to \psi) \to (\varphi \to \psi) \equiv (\varphi \to \psi) \to \neg\neg(\varphi \to \psi) \quad (dn)$$
$$\equiv (\varphi \to \psi) \to \neg(\neg\varphi \to \psi) \quad (2.4.19)$$

$\square$

In Section 4.3.1 there is another way of negating the conditional according to
$$\vdash \neg(\varphi \to \psi) \equiv (\varphi \to \neg\psi) \vee (\neg\varphi \to \psi)$$

Yet another way of negating the conditional, modalizing the negated consequent, is mentioned in Section 7.2.

2.5 Rejecting classical validities

Since Connexive logics are contra-classical, obviously there are some classical validities rejected by them. An impetus for some of the rejections is the following proposition.

Proposition 2.5.2 (incompatibility with connexivity). *If a logic $\mathcal{L}$ with a conditional '$\to$', a negation '$\neg$' and a conjunction '$\wedge$' satisfies the following classical properties:*

- *Modus ponens (MP) as a rule:*

$$\dfrac{\varphi \to \psi \quad \varphi}{\psi} \ (MP) \tag{2.5.23}$$

- *Conjunction simplification:*

$$\vdash_\mathcal{L} \varphi \wedge \psi \to \varphi \qquad \vdash_\mathcal{L} \varphi \wedge \psi \to \psi \tag{2.5.24}$$

- *Contraposition:*

$$\vdash_\mathcal{L} (\varphi \to \psi) \to (\neg \psi \to \neg \varphi) \tag{2.5.25}$$

- *Transitivity of '$\to$':*

$$\vdash_\mathcal{L} ((\varphi \to \psi) \wedge (\psi \to \chi)) \to (\varphi \to \chi) \tag{2.5.26}$$

or, for convenience, as a rule

$$\dfrac{\varphi \to \psi \quad \psi \to \chi}{\varphi \to \chi} \ (\to TR)$$

then $\mathcal{L}$ is not a Connexive logic.

Proof. The following is an $\mathcal{L}$-derivation the consequence of which contradicts A_1.

1. $\varphi \wedge \neg \varphi \to \varphi$ *Conjunction elimination*
2. $\varphi \wedge \neg \varphi \to \neg \varphi$ *Conjunction elimination*
3. $((\varphi \wedge \neg \varphi) \to \varphi) \to (\neg \varphi \to \neg(\varphi \wedge \neg \varphi))$ *contraposition*
4. $\neg \varphi \to \neg(\varphi \wedge \neg \varphi)$ 3., 1., MP
5. $\varphi \wedge \neg \varphi \to \neg(\varphi \wedge \neg \varphi)$ 2., 4., (TR)

□

2.5 Rejecting classical validities

A similar proposition holds for disjunction and the classical laws of disjunction addition:

$$\vdash_{\mathcal{L}} \varphi \to (\varphi \lor \psi) \quad \vdash_{\mathcal{L}} \psi \to (\varphi \lor \psi) \tag{2.5.27}$$

In this section, I delineate the main rejected classical laws. Later, for each Connexive logic presented in detail, some specific classical validities rejected by that logic are specified.

2.5.1 Rejecting conjunction simplification

As mentioned above, the following two laws of Classical logic are known as *conjunction simplification* (and also as conditional simplification):

$$(Conj_1)\ \varphi \land \psi \supset \varphi \quad (Conj_2)\ \varphi \land \psi \supset \psi \tag{2.5.28}$$

The same property of conjunction appears also as the conjunction elimination rules in natural-deduction systems:

$$\frac{\varphi \land \psi}{\varphi}\ (\land E_1) \quad \frac{\varphi \land \psi}{\psi}\ (\land E_2) \tag{2.5.29}$$

Both formulations strongly reflect the truth-functionality of the conjunction in Classical logic.

By taking the conditional in (2.5.28) to be a connexive '→' and by taking ψ as $\neg\varphi$, we get by the simplification laws another incompatibility of conjunction simplification with connexivity, in addition to Proposition 2.5.2.

$$\vdash (\varphi \land \neg\varphi \to \varphi) \land (\varphi \land \neg\varphi \to \neg\varphi)$$

which has the form

$$\vdash (\chi \to \varphi) \land (\chi \to \neg\varphi)$$

Contradicting B_1.
So, when conjunctive simplification is incorporated with connexivity it

may generate contradictions. See, for example, Wansing's logic C in Section 3.2, which validates conjunctive simplification and is negation inconsistent.

Also at the level of logical consequence, the following is a facet of the same rejection:

$$\varphi \wedge \psi \not\vdash \varphi \qquad \varphi \wedge \psi \not\vdash \psi \qquad (2.5.30)$$

In particular, some paraconsistent logics reject *any* logical consequences out of a contradiction. In particular,

$$\varphi \wedge \neg\varphi \not\vdash \varphi \qquad \varphi \wedge \neg\varphi \not\vdash \neg\varphi$$

Several arguments were raised in the literature, defending the rejection of conjunctive simplification. Nelson [79] raises arguments of *irrelevance*: it is not the whole conjunction that implies a conjunct – merely that conjunct itself, the other being non-relevant to the implication, not contributing to it. Nelson attributes the failure for conjunctive simplification for the non-truth-functional multiplicative conjunction (fusion) to the same reason.

In a paper dedicated to the defense of the failure of conjunctive simplification, Thompson [116] observes the following "strange" inference via conjunctive simplification:

$$\vdash \varphi \wedge \neg(\varphi \rightarrow \varphi) \rightarrow \varphi$$

This formula is read by Thompson as the assertion that φ follows from φ, even under the condition that it does not.

2.5.2 Rejecting disjunction addition

The following two laws of Classical logic are known as *disjunction addition*.

$$(Dis_1)\ \varphi \rightarrow (\varphi \vee \psi) \qquad (Dis_2)\ \psi \rightarrow (\varphi \vee \psi) \qquad (2.5.31)$$

2.5 Rejecting classical validities

The same property of disjunction is captured also by the disjunction introduction rule in natural-deduction systems:

$$\frac{\varphi}{\varphi \vee \psi} \; (\vee I_1) \quad \frac{\psi}{\varphi \vee \psi} \; (\vee I_1) \tag{2.5.32}$$

By instantiating ψ to $\neg\varphi$ in (Dis_2), we get that $\varphi \vee \neg\varphi$ is implied both by φ and by $\neg\varphi$, contradicting B_3, B_4 and AB_2; hence, t disjunction addition, when combined with with connexivity, may also lead to negation inconsistency.

A family of logics that systematically invalidate disjunctive addition is the family of *containment logics* based on a principle known as *the proscriptive principle*:

$$(PP): \; \vdash \varphi \rightarrow \psi \text{ iff } At[\![\psi]\!] \subseteq At[\![\varphi]\!] \tag{2.5.33}$$

That is, in a valid conditional all the atoms of the consequent are included in the antecedent. The (PP) principle implements one variant of *relevance* of the consequent to the antecedent in a valid conditional - a topic to which I return in the sequel. For a detailed discussion and study of this principle, see Ferguson [27, 31].

Obviously, disjunctive addition (in either of its two forms above) does not adhere to (PP), as ψ may add atoms not included in φ.

Yet another reason to reject disjunctive addition can be rooted in *Pragmatics*. In his work on the *cooperative* aspects of natural language use, Grice [42, 43] sets up *maxims* that drive conversation. One of the maxims is that of *quantity*, where one is required to be as informative as one possibly can, and give as much information as is needed, and no more. The rules of disjunction addition can be seen as violating this maxim: if a speaker asserts a disjunction $\varphi \vee \psi$ while already knowing one of the disjuncts, the speaker gives less information than the speaker possesses. For a discussion of this (and related matters) see Terrés Villalonga [118].

2.5.3 Rejecting contraposition

As mentioned above, the following law of Classical logic is known as *contraposition*

$$\text{For every } \varphi \text{ and } \psi: \quad \vdash (\varphi \to \psi) \to (\neg\psi \to \neg\varphi) \tag{2.5.34}$$

It also appears as a property of logical consequence:

$$\text{For every } \varphi \text{ and } \psi: \quad \varphi \to \psi \vDash \neg\psi \to \neg\varphi \tag{2.5.35}$$

This law also strongly depends on the truth-functionality of the classical material conditional.

Contraposition is rejected in Nelson's constructive logics with strong negation $N3$ and $N4$. As Kapsner [55] (p. 155) explains, the reason is that verification and falsification in these logics are loosely connected, merely preventing both the verification and the falsification of the same proposition.

A good reason to reject contraposition is the distinction between truth and non-falsity, as well as falsity and non-truth, distinctions playing an important role in the definition of some Connexive logics.

In logics in which falsity and non-truth come apart, like Kleene's three-valued logic K_3, Priest's three-valued logic of paradox LP and the Belnap-Dunn four-valued logic FDE (see Section 7.4.1.2 for the latter), contraposition fails.

In such logics, acceptable conditionals may have non-truth-apt antecedents and false consequents; but their contrapositives, which have true antecedents and non-truth-apt consequents, are not acceptable.

Note that the converse form of contraposition, $(\neg\psi \to \neg\varphi) \to (\varphi \to \psi)$ is rejected by Intuitionistic logic, as it leads to the validity of the double-negation elimination.

2.6 Objections to the connexive characteristics

In spite of the plausibility of the connexive principles as properties of the NL conditional, there were some explicit objections to those characteristic axioms (beyond restricting their scope, the latter discussed in Chapter 4).

The common theme of those objections is the conflict between the connexive axioms with other well entrenched classical principles. So, it seems that those objections are more objections to the contra-classicality of Connexive logics than objections to the connexive axioms per se.

- In [70], Martin, deeply embedded within a description of 12th century theories of the conditional, mentions an objection to Aristotle's axioms, because they lead to contradictions together with some commonly accepted principles the conditional is seen to obey. Martin formulates the connexive axiom as

$$(AR) \vdash \neg((\varphi \to \psi) \land (\neg \varphi \to \neg \psi)) \qquad (2.6.36)$$

Subsequently, he presents the following derivation of a contradiction (notationally adjusted), which he coins "unfortunate":

1. $\varphi \land \neg \varphi \to \varphi$ [$conjunction\ simplification$]

2. $\varphi \land \neg \varphi \to \neg \varphi$ [$conjunction\ simplification$]

3. $\neg \varphi \to \neg(\varphi \land \neg \varphi)$ [1., $contraposition$]

4. $\varphi \to \neg(\varphi \land \neg \varphi)$ [2., $double\ negation,\ contraposition$]

5. $((\varphi \to \neg(\varphi \land \neg \varphi)) \land (\neg \varphi \to \neg(\varphi \land \neg \varphi)))$ [3., 4., $conjunction$]

6. $\neg((\varphi \to \neg(\varphi \land \neg \varphi)) \land (\neg \varphi \to \neg(\varphi \land \neg \varphi)))$ [AR, $\varphi \land \neg \varphi / \psi$]

$$(2.6.37)$$

However, such criticism is somewhat of begging the question because of the contra-classicality of Connexive logics, giving up some classical properties of the conditional. Recall the rejection of conjunction simplification considered in Section 2.5.1.

As Martin indicates, for example, McCall's Connexive logic (see Section 3.6.2) also gives up conditional simplification.

- In [64], (p. 50-51) Łukasiewicz attempts to refute Aristotle's axiom by showing it to contradict an *arithmetical* (– not a logical!) principle, used by Euclid, as follows.

    > [Euclid] states first that 'If the product of two integers, a and b, is divisible by a prime number n, then if a is not divisible by n, b should be divisible by n.' Let us now suppose that $a = b$ and the product $a \times a$ (a^2) is divisible by n. It results from this supposition that 'If a is not divisible by n, then a is divisible by n.' Here we have an example of a true conditional the antecedent of which is the negation of the consequent.

    However, as explained by Ferguson [30],

    > Despite Euclid's employing a conditional in the statement of the lemma, to state that if a prime n divides a composite number $a \times b$ then either n divides a or n divides b is an equally good (and perhaps superior) formulation of Euclid's lemma. If so, Łukasiewicz' argument requires an enthymematic assumption of the validity of disjunctive syllogism.

    So, the conflict is once again between the connexive axioms and a classical principle.

- In [105], Routley and Montgomery object to the connexive axioms on the grounds that adding them to several formal systems that contain

[a] number of principles of implication which it is very difficult to reject on semantic grounds

yields inconsistencies. Again, the same conflict with classical principles emerges. See Bode [12] for a lengthy refutation of Routley and Montgomery's arguments.

- As reported by Mares and Paoli [68], C. I. Lewis rejects the connexive axioms. He does so in an objection implicitly directed at Nelson's logic, and again because the invalidation of classical principles, in particular conjunction simplification.

2.7 Empirical corroboration of connexivity

In this section, I review some empirical studies that show that people have a disposition to interpret NL conditionals in a way compatible with connexivity. More on other empirical tests of connexivity can be found in Wansing [122].

In [74, Section 2], McCall reports some results on testing the endorsement of connexive principles given by indicative conditionals in English in concrete form on a group of 89 non-expert philosophy students at McGill University in Canada.

The following is an example sentence, which the students had to judge for truth-value.

If Hitler is dead then (if Hitler is not dead then Hitler is dead).

In addition, the participants in the experiment where asked to judge short arguments for validity. For example,

If the fires are lit the battle is over.
Therefore, it is false that if the fires are lit the battle is not over.

The findings of the experiment support the intuition that laymen speakers of English subscribe to those connexive principles to a rather high degree: 88% in the case of A_1 and 84% in the case of B_1.

Empirical studies on A_1 and A_2 have been carried out by Pfeifer [89]. In one experiment, the sample consisted of 141 psychology students (110 females and 31 males) at the University of Salzburg, Austria. Both A_1 and A_2 were tested as abstract as well as concrete indicative conditionals. In a second experiment, 40 students without training in logic (20 females and 20 males) had to solve tasks involving concrete indicative conditionals in English. In this case, scope ambiguities arising from the negation of conditionals were ruled out. Both experiments provide evidence against the interpretation of indicative conditionals in English as material conditional and support the connexive reading of negated conditional expressed by Aristotle's theses.

Pfeifer sees these findings as strong evidence for interpreting indicative conditionals as conditional events. This interpretation predicts that people should strongly believe that Aristotle's theses are valid because the only coherent assessment for them is the probability value 1.

Chapter 3

A sample of connexive logics

In this chapter, I review in some detail a sample of propositional Connexive logics presented in the literature. This is by no means an exhaustive sample. It is merely intended to give the reader a grasp of what is involved in devising Connexive logics in a principled way. Furthermore, it puts emphasis on those Connexive logics more closely related to my own work. More Connexive logics, as well as as more details on the logics surveyed in this chapter, may be found in Wansing [122].

For each Connexive logic in the sample, after its presentation, I indicate:

- Proofs of the connexive axioms in that logic.

- The way the conditional is negated in that logic.

- Proofs of paraconsistency and paracompleteness in that logic.

- Some classical laws/rules given up by that logic.

Where available, 1st-order extensions of the logic are presented too, in a later chapter.

At the end of the chapter I also present briefly, without full details, several more Connexive logics found in the literature, in order to exemplify the wealth and variety of this topic.

3.1 How are connexive logics obtained?

A natural question in this context is the following: *how* are Connexive logics obtained? What brings about the validity of the connexive axioms?

Several methods for obtaining connexivity emerged in the literature.

- In an axiomatic definition of the logic, add one (or more) of the connexive characteristics as part of the definition. For example, in Wansing's Connexive logic C (see Section 3.2.1), the axiom expressing negating the conditional is added to the positive fragment of Intuitionistic logic.

$$\vdash_C \sim(\varphi\to\psi) \leftrightarrow (\varphi\to\sim\psi)$$

  In Angell's Connexive logic $\mathbf{PA_1}$ (see Section 3.4.2) B_1 is included as an axiom.

$$\vdash_{PA_1} (\varphi\to\psi)\to\neg(\varphi\to\neg\psi)$$

- In a model-theoretic definition of the logic, modify the falsification condition of the conditional. This is a general method for obtaining contra-classicality in general, and connexivity in particular, suggested by Wansing. In Estrada-González [24], this approach was coined as the *Bochum plan*, and is studied in details, with many examples of contra-classical logics, including Connexive logics, generated in this way.

According to Omori's formulation in [85, Section 1.1], Wansing suggests to take the condition of the form "*if φ is true then ψ is false*" rather than the (classical) condition of the form "*φ is true and ψ is false*" as the falsification condition for the conditional of the form "if φ then ψ", where in Wansing's proposal support of truth and support of falsity are not necessarily mutually exclusive nor exhaustive.

- When defining a Connexive logic via the logical consequence relation, strengthen the definition of logical consequence. For example, in Priest's Connexive logic P_S (see Section 3.3), logical consequence is strengthened so as to *avoid vacuity*: for $\Gamma \vDash \varphi$ to hold, Γ has to be satisfiable, as well as $\neg\varphi$. Consequently, the verification condition of the conditional, internalizing logical consequence, is modified accordingly.

- In a definition using natural-deduction I/E-rules are used for the negated conditional. See the definition in Section 3.5.

- There is also a way to obtain connexivity by an appeal to a cut free sequent calculus by McCall [73], where every theorem is obtained from two non-logical axioms (initial sequents).

Some more methods are presented at the end of the section.

3.2 Wansing's propositional Connexive logic C

As the first propositional Connexive logic to present I take Wansing's propositional Connexive logic C as introduced in [119]. Its axiomatic definition is presented in Section 3.2.1, and its connexivity is proven axiomatically in Section 3.2.2. Its model-theoretic specification is presented in Section 3.2.3, and the connexivity of C is shown model-theoretically in Section 3.2.4. The paraconsistency of C is shown in Section 3.2.5 and some classical validities invalidated by C are shown in Section 3.2.6. Some variations on C are presented in Section 3.2.7. Its first-order extension QC is considered in Section 6.5.

3.2.1 Axiomatic presentation of C

The object-language is the same as that for propositional Classical logic, namely the closure of $\mathcal{P}$, the atomic propositions, under '$\wedge$', '$\vee$', '$\rightarrow$', '$\sim$', with a connexive conditional '$\rightarrow$' taking the place of the material implication '$\supset$', and a strong negation '$\sim$' replaces the boolean negation '$\neg$'.

The connexive biconditional '$\leftrightarrow$' is defined as $(\varphi \rightarrow \psi) \wedge (\psi \rightarrow \varphi)$.

The logic is an extension of positive intuitionistic propositional logic, all the theses[1] of which are considered as axioms.

$$
\begin{array}{ll}
\varphi \rightarrow (\psi \rightarrow \varphi) & (Ax1) \\[4pt]
(\varphi \rightarrow (\psi \rightarrow \chi)) \rightarrow ((\varphi \rightarrow \psi) \rightarrow \varphi \rightarrow \chi) & (Ax2) \\[4pt]
\varphi \wedge \psi \rightarrow \varphi & (Ax3) \\[4pt]
\varphi \wedge \psi \rightarrow \psi & (Ax4) \\[4pt]
((\chi \rightarrow \varphi) \rightarrow ((\chi \rightarrow \psi) \rightarrow (\chi \rightarrow \varphi \wedge \psi))) & (Ax5) \\[4pt]
\varphi \rightarrow (\varphi \vee \psi) & (Ax6) \\[4pt]
\psi \rightarrow (\varphi \vee \psi) & (Ax7) \\[4pt]
(\varphi \rightarrow \chi) \rightarrow ((\psi \rightarrow \chi) \rightarrow ((\varphi \vee \psi) \rightarrow \chi)) & (Ax8)
\end{array}
\qquad (3.2.1)
$$

---

[1] In derivations, an appeal to an intuitionistic thesis is justified as (Int).

3.2 Wansing's propositional Connexive logic C

In addition, we have the following negative axioms, with added mnemonic names for use in examples.

$$\sim\sim \varphi \leftrightarrow \varphi \qquad (Ax9) - (DN)$$
$$\sim (\varphi \vee \psi) \leftrightarrow (\sim \varphi \wedge \sim \psi) \quad (Ax10) - (DM1)$$
$$\sim (\varphi \wedge \psi) \leftrightarrow (\sim \varphi \vee \sim \psi) \quad (Ax11) - (DM2)$$
$$\sim (\varphi \rightarrow \psi) \leftrightarrow (\varphi \rightarrow \sim \psi) \quad (Ax12) - (NC)$$

(3.2.2)

There is one inference rule, modus-ponens (MP).

$$\frac{\varphi \rightarrow \psi \quad \varphi}{\psi} \ (MP)$$

Clearly, the axiom (NC), that of negating the conditional, is the source of connexivity, and of contra-classicality of C. It embodies a falsification condition for the connexive conditional, rendering the latter different from the classical one, which is defined as $\sim \varphi \vee \psi$ and negated as $\varphi \wedge \sim \psi$.

3.2.2 Axiomatic derivability of the connexive axioms

Proposition 3.2.3 (proving the connexive axioms). *The propositional connexive axioms are provable in C.*

A_1:

(1) $(\varphi \rightarrow \sim\sim \varphi) \rightarrow \sim (\varphi \rightarrow \sim \varphi)$ (NC)

(2) $\varphi \rightarrow \sim\sim \varphi$ (DN)

(3) $\sim (\varphi \rightarrow \sim \varphi)$ $(1), (2), MP$

(3.2.3)

A_2:

$$(1)\ (\sim\varphi \to \sim\varphi) \to \sim(\sim\varphi \to \varphi)\quad (NC)$$

$$(2)\ \sim\varphi \to \sim\varphi \qquad (Int)$$

$$(3)\ \sim(\sim\varphi \to \varphi)\qquad (1),(2), MP$$

(3.2.4)

B_1: An instance of (NC) (right to left) with ψ taken as $\sim\psi$ and using (DN).

B_2: Similar.

3.2.3 Complete models for C

The following specification of models for C is taken from Wansing [119]. The two main features of the connexive model-theory are the following:

1. Present independently separate *verification conditions* and *falsification conditions*, as in Nelson's logic **N4** [78] for strong (constructive) negation.

2. Let both verification and falsification conditions for the connexive conditional be *dynamic*, i.e., depending on more than the point of evaluation.

A C-frame is a tuple $\mathcal{F} = \langle W, \leq \rangle$, where W is a non-empty set of *points* (of evaluation) ranged over by s, t and '$\leq$' a reflexive and transitive binary relation on W.

A C-model $\mathcal{M}$ is a tuple $\mathcal{M} = \langle W, \leq, v^+, v^- \rangle$, where:

- $\langle W, \leq \rangle$ is a C-frame.

3.2 Wansing's propositional Connexive logic C

- v^+ and v^- map atomic formulas to subsets of W closed under $\leq$: if $t \in v^+[\![p]\!]$ and $t \leq s$, then $t \in v^-[\![p]\!]$ (persistence[2]; similarly for v^-.

The two relations $\models^+$ (support of truth) and $\models^-$ (support of falsity) are jointly defined inductively as follows.

$$\mathcal{M}, t \models^+ p \quad \text{iff} \quad t \in v^+(p)$$

$$\mathcal{M}, t \models^- p \quad \text{iff} \quad t \in v^-(p)$$

$$\mathcal{M}, t \models^+ \varphi \wedge \psi \quad \text{iff} \quad \mathcal{M}, t \models^+ \varphi \text{ and } \mathcal{M}, t \models^+ \psi$$

$$\mathcal{M}, t \models^- \varphi \wedge \psi \quad \text{iff} \quad \mathcal{M}, t \models^- \varphi \text{ or } \mathcal{M}, t \models^- \psi$$

$$\mathcal{M}, t \models^+ \varphi \vee \psi \quad \text{iff} \quad \mathcal{M}, t \models^+ \varphi \text{ or } \mathcal{M}, t \models^+ \psi \quad (3.2.5)$$

$$\mathcal{M}, t \models^- \varphi \vee \psi \quad \text{iff} \quad \mathcal{M}, t \models^- \varphi \text{ and } \mathcal{M}, t \models^- \psi$$

$$\mathcal{M}, t \models^+ \varphi \to \psi \quad \text{iff} \quad \forall s \geq t: \text{ if } \mathcal{M}, s \models^+ \varphi \text{ then } \mathcal{M}, s \models^+ \psi$$

$$\mathcal{M}, t \models^- \varphi \to \psi \quad \text{iff} \quad \forall s \geq t: \text{ if } \mathcal{M}, s \models^+ \varphi \text{ then } \mathcal{M}, s \models^- \psi$$

$$\mathcal{M}, t \models^+ \sim\varphi \quad \text{iff} \quad \mathcal{M}, t \models^- \varphi$$

$$\mathcal{M}, t \models^- \sim\varphi \quad \text{iff} \quad \mathcal{M}, t \models^+ \varphi$$

As for the connexive biconditional, we get the following clauses:

$$\mathcal{M}, t \models^+ \varphi \leftrightarrow \psi \quad \text{iff} \quad \forall s \geq t: \mathcal{M}, s \models^+ \varphi \text{ iff } \mathcal{M}, s \models^+ \psi$$

$$\mathcal{M}, t \models^- \varphi \leftrightarrow \psi \quad \text{iff} \quad \forall s \geq t: \begin{array}{c} \text{if } \mathcal{M}, s \models^+ \varphi \text{ then } \mathcal{M}, s \models^- \psi \\ \text{and} \\ \text{if } \mathcal{M}, s \models^+ \psi \text{ then } \mathcal{M}, s \models^- \varphi \end{array} \quad (3.2.6)$$

---

[2] It can be shown that persistent via support of truth/falsity extends to all formulas.

Definition 3.2.1 (validity). *There are three levels of validity:*

1. φ *is valid in a model* $\mathcal{M}$, *denoted* $\models_{\mathcal{M}}\varphi$, *iff* $\mathcal{M}, t \models^{+} \varphi$ *for every* $t \in W$.

2. φ *is valid in a frame* $\mathcal{F}$, *denoted* $\models_{\mathcal{F}}\varphi$, *iff* $\models_{\mathcal{M}}\varphi$ *for every model* $\mathcal{M}$ *based on* $\mathcal{F}$.

3. φ *is C-valid, denoted* $\models_{C}\varphi$, *iff it is valid for every frame* $\mathcal{F}$.

Logical consequence, $\Gamma \models_{C} \varphi$, is defined similarly.

For a discussion and proof of the completeness of C-models, see Wansing [119].

Digression: the informational interpretation of frames

Wansing offers an intuitive interpretation of the frames above in terms of *information*. The points of evaluation represent *states of information*, that are capable of supporting the truth (giving evidence in favor) or falsity (giving evidence against) propositions. The '≤' relation is viewed as expansion of states to more informative ones. States of information need not be consistent, and can support both the truth and the falsity of some propositions. Neither need states of information be complete, and may provide no information (i.e., have no evidence regarding that proposition). This is a major difference from Kripke frames for modeling modality, the latter employing *worlds*: consistent and total states.
(end of digression)

3.2.4 Model-theoretic establishment of the connexive axioms

To see the effect of the dynamic falsification condition, which is responsible for obtaining a connexive conditional, consider the establish-

3.2 Wansing's propositional Connexive logic C

ment of the validity in C of the connexive axioms, this time model-theoretically. I show two of them.

A_1:

$\models_C \sim (\varphi \to \sim \varphi)$ iff $(\forall \mathcal{F}) \models_\mathcal{F} \sim (\varphi \to \sim \varphi)$

$\phantom{\models_C \sim (\varphi \to \sim \varphi) \text{ iff }}$ iff $(\forall \mathcal{M} \text{ based on } \mathcal{F}) \models_\mathcal{M} \sim (\varphi \to \sim \varphi)$

$\phantom{\models_C \sim (\varphi \to \sim \varphi) \text{ iff }}$ iff $(\forall t \in W)\, \mathcal{M}, t \models^+ \sim (\varphi \to \sim \varphi)$

$\phantom{\models_C \sim (\varphi \to \sim \varphi) \text{ iff }}$ iff $(\forall t \in W)\, \mathcal{M}, t \models^- (\varphi \to \sim \varphi)$

$\phantom{\models_C \sim (\varphi \to \sim \varphi) \text{ iff }}$ iff $(\forall t \in W)(\forall s \geq t):$ if $\mathcal{M}, s \models^+ \varphi$ then $\mathcal{M}, s \models^- \sim \varphi$

$\phantom{\models_C \sim (\varphi \to \sim \varphi) \text{ iff }}$ iff $(\forall t \in W)(\forall s \geq t):$ if $\mathcal{M}, s \models^+ \varphi$ then $\mathcal{M}, s \models^+ \varphi$

which clearly holds.

(3.2.7)

B_1:

$\models_C (\varphi \to \psi) \to \sim (\varphi \to \sim \psi)$

iff $(\forall \mathcal{F}) \models_{\mathcal{F}} (\varphi \to \psi) \to \sim (\varphi \to \sim \psi)$

iff $(\forall \mathcal{M} \text{ based on } \mathcal{F}) \models_{\mathcal{M}} (\varphi \to \psi) \to \sim (\varphi \to \sim \psi)$

iff $(\forall t \in W)\ \mathcal{M}, t \models^+ (\varphi \to \psi) \to \sim (\varphi \to \sim \psi)$

iff $(\forall t \in W)(\forall s \geq t)$ if $\mathcal{M}, s \models^+ (\varphi \to \psi)$ then $\mathcal{M}, s \models^+ \sim (\varphi \to \sim \psi)$

iff $(\forall t \in W)(\forall s \geq t)$ if $\mathcal{M}, s \models^+ (\varphi \to \psi)$ then $\mathcal{M}, s \models^- \varphi \to \sim \psi$

iff $(\forall t \in W)(\forall s \geq t)$ if
  $[(\forall r \geq s)$ if $\mathcal{M}, r \models^+ \varphi$ then $\mathcal{M}, r \models^+ \psi]$
  then
  $(*)\ [(\forall r \geq s)$ if $\mathcal{M}, r \models^+ \varphi$ then $\mathcal{M}, r \models^- \sim \psi]$

iff $(\forall t \in W)(\forall s \geq t)$ if
  $[(\forall r \geq s)$ if $\mathcal{M}, r \models^+ \varphi$ then $\mathcal{M}, r \models^+ \psi]$
  then
  $(*)\ [(\forall r \geq s)$ if $\mathcal{M}, r \models^+ \varphi$ then $\mathcal{M}, r \models^+ \psi]$

which clearly holds.

The other axioms are shown similarly.

3.2 Wansing's propositional Connexive logic C

In C, B_1', the converse of B_1, is also provable.

$\models_C \sim (\varphi \to \psi) \to (\varphi \to \sim \psi)$ iff $(\forall \mathcal{F}) \models_{\mathcal{F}} \sim (\varphi \to \psi) \to (\varphi \to \sim \psi)$

iff $(\forall \mathcal{M} \text{ based on } \mathcal{F}) \models_{\mathcal{M}}$
$\sim (\varphi \to \psi) \to (\varphi \to \sim \psi)$

iff $(\forall t \in W)\, \mathcal{M}, t \models^+ \sim (\varphi \to \psi) \to (\varphi \to \sim \psi)$

iff $(\forall t \in W)(\forall s \geq t)$ if $\mathcal{M}, s \models^+ \sim (\varphi \to \psi)$
then $\mathcal{M}, s \models^+ (\varphi \to \sim \psi)$

iff $(\forall t \in W)(\forall s \geq t)$ if $\mathcal{M}, s \models^- (\varphi \to \psi)$
then $\mathcal{M}, s \models^+ (\varphi \to \sim \psi)$

iff $(\forall t \in W)(\forall s \geq t)$ if
$[(\forall r \geq s) \text{ if } \mathcal{M}, r \models^+ \varphi \text{ then } \mathcal{M}, r \models^- \psi]$
then
$(*)\ [(\forall r \geq s) \text{ if } \mathcal{M}, r \models^+ \varphi \text{ then } \mathcal{M}, r \models^+ \sim \psi]$

iff $(\forall t \in W)(\forall s \geq t)$ if
$[(\forall r \geq s) \text{ if } \mathcal{M}, r \models^+ \varphi \text{ then } \mathcal{M}, r \models^- \psi]$
then
$(*)\ [(\forall r \geq s) \text{ if } \mathcal{M}, r \models^+ \varphi \text{ then } \mathcal{M}, r \models^- \psi]$

which clearly holds.

Example 3.2.5 (Counter-model to Abelard AB_1). *Perhaps surprisingly, we have that C invalidates Abelard's connexive axioms:*

$$\not\models_C \sim ((\varphi \to \psi) \wedge (\varphi \to \sim \psi)) \tag{3.2.8}$$

For a counter-model for AB_1, let $W = \{t, s\}$ with $s \leq t$, and $(*)\ v^+[\![p]\!] = \{t, s\}$ and $(**)\ v^-[\![p]\!] = v^-[\![q]\!] = \emptyset$.
¿From $(*)$: $\mathcal{M}, t \models^+ p$ and $\mathcal{M}, s \models^+ p$,
and from $(**)$: $\mathcal{M}, t \not\models^- q$ and $\mathcal{M}, s \not\models^- q$, as well as $\mathcal{M}, t \not\models^+ q$ and

$\mathcal{M}, s \not\models^+ q$.

Therefore, $\mathcal{M}, t \not\models^- p \to q$ and $\mathcal{M}, t \not\models^- p \to \sim q$, implying $(***)$ $\mathcal{M}, t \not\models^- p \to q \wedge p \to \sim q$. Finally, $(***)$ implies the required $\mathcal{M}, t \not\models^+ \sim (p \to q \wedge p \to \sim q)$, establishing the invalidity of AB_1.

□

Proposition 3.2.4 (Unprovability of $(asym)$).

Proof. As a counter-model for $(\varphi \to \psi) \to (\psi \to \varphi)$ $(asym)$ in (2.2.12)), consider the following model $\mathcal{M}_{asym}$. For some atom p, let $v^+[\![p]\!] = \emptyset$ and $v^-[\![p]\!] = W$. We get:

- For no $t \in W$: $\mathcal{M}, t \models^+ p \wedge \sim p$.

- For every $t \in W$: $\mathcal{M}, t \models^+ p \vee \sim p$.

Consequently:

- For every $t \in W$: $\mathcal{M}, t \models^+ p \wedge \sim p \to p \vee \sim p$.

- For no $t \in W$: $\mathcal{M}, t \models^+ p \vee \sim p \to p \wedge \sim p$.

Hence,
$$\not\models_C (p \wedge \sim p \to p \vee \sim p) \to (p \vee \sim p \to p \wedge \sim p)$$
where the latter is an instance of $(\varphi \to \psi) \to (\psi \to \varphi)$. □

Proposition 3.2.5 (C is not strongly-connexive). *$(\varphi \wedge \sim \varphi) \to \sim (\varphi \wedge \sim \varphi)$ is satisfiable (cf. Section 2.2.0.2).*

3.2 Wansing's propositional Connexive logic C

Proof. A model $\mathcal{M}$ or $(\varphi \wedge {\sim} \varphi) \to {\sim}(\varphi \wedge {\sim}\varphi)$ (an instance of $\varphi \to {\sim}\varphi$) is constructed as follows. $W = \{t\}$, $t \leq t$ and $t \in v^+[\![p]\!] \cap v^-[\![p]\!]$. Hence

$$(*)\ \mathcal{M}, t \vDash^+ p \text{ and } (**)\ \mathcal{M}, t \vDash^- p$$

From $(**)$, $\mathcal{M}, t \vDash^+ {\sim}p$. Therefore, $\mathcal{M}, t \vDash^+ p \wedge {\sim}p$ and $\mathcal{M}, t \vDash^+ {\sim}(p \wedge {\sim}p)$, implying the required $\mathcal{M}, t \vDash^+ (p \wedge {\sim}p) \to {\sim}(p \wedge {\sim}p)$.

The other satisfiabilities are proved similarly. □

3.2.5 Paraconsistency and paracompleteness of C

Proposition 3.2.6 (negation inconsistency of C). *C is (negation-) inconsistent.*

Proof. The following two contradictory formulas are both C-valid.

$$(i)\ \vDash_C \varphi \wedge {\sim}\varphi \to \varphi \qquad (ii)\ \vDash_C {\sim}(\varphi \wedge {\sim}\varphi \to \varphi)$$

A model-theoretic proof: Consider any frame $\mathcal{F}$ and any model $\mathcal{M}$ based on $\mathcal{F}$.

ad (i):

$$(\forall t \in W)\ \mathcal{M}, t \vDash^+ \varphi \wedge {\sim}\varphi \to \varphi$$
iff
$$(\forall t \in W)(\forall s \geq t) \text{ if } \mathcal{M}, s \vDash^+ \varphi \wedge {\sim}\varphi \text{ then } \mathcal{M}, s \vDash^+ \varphi$$

The latter clearly holds, since

$$\mathcal{M}, s \vDash^+ \varphi \wedge {\sim}\varphi \text{ iff } \mathcal{M}, s \vDash^+ \varphi \text{ and } \mathcal{M}, s \vDash^+ {\sim}\varphi$$

ad (ii):

$(\forall t \in W)\ \mathcal{M}, t \vDash^+ {\sim}(\varphi \wedge {\sim}\varphi \to \varphi)$ iff $\mathcal{M}, t \vDash^- \varphi \wedge {\sim}\varphi \to \varphi$
iff $(\forall t \in W)(\forall s \geq t)$ if $\mathcal{M}, s \vDash^+ \varphi \wedge {\sim}\varphi$ then $\mathcal{M}, s \vDash^- \varphi$

The latter clearly holds, since

$$\mathcal{M}, s \models^+ \varphi \wedge \sim \varphi \text{ iff } \mathcal{M}, s \models^+ \varphi \text{ and } \mathcal{M}, s \models^- \varphi$$

An axiomatic proof:

1.	$\varphi \wedge \sim \varphi \to \sim \varphi$	$Ax\ 4$
2.	$(\varphi \wedge \sim \varphi \to \sim \varphi) \to \sim (\varphi \wedge \sim \varphi \to \varphi)$	$Ax\ 12$
3.	$\sim (\varphi \wedge \sim \varphi \to \varphi)$	1., 2., (MP)

contradicting $(Ax3)$.

The following contradiction to A_1 is derivable axiomatically in a logic extending Anderson and Belnap's weak Relevance logic B with A_1, as shown by Mortensen [75]. However, the derivation carries also to C.

1.	$\varphi \wedge \sim \varphi \to \varphi$	conjunction simplification
2.	$\sim \varphi \to (\sim \varphi \vee \sim\sim \varphi)$	disjunction addition
3.	$(\varphi \wedge \sim \varphi) \to (\sim \varphi \vee \sim\sim \varphi)$	1., 2., transitivity
4.	$(\sim \varphi \vee \sim\sim \varphi) \to (\sim \varphi \vee \sim\sim \varphi)$	identity
5.	$(\sim \varphi \vee \sim\sim \varphi) \to (\varphi \wedge \sim \varphi)$	4., De Morgan
6.	$(\varphi \wedge \sim \varphi) \to \sim (\varphi \wedge \sim \varphi)$	3., 5., transitivity

$\square$

However, the presence of contradictions does not trivialize C, as follows from the next proposition.

Proposition 3.2.7 (paraconsistency and paracompleteness of C). *C is paraconsistent and paracomplete.*

Proof. 1. Consider a model $\mathcal{M}$, atoms p, q, and a valuation v where for $t \in W$ we have both $t \in v^+(p)$ and $t \in v^-(p)$, as well as $t \notin v^+(q)$.

Thus, $\mathcal{M}, t \models^+ p \wedge \sim p$, but $\mathcal{M}, t \not\models^+ q$, establishing $p \wedge \sim p \not\models_C q$, hence the paraconsistency of C.

2. Similarly, consider a model $\mathcal{M}$, atoms p, q, and a valuation v where for $t \in W$ we have both $t \notin v^+(q)$ and $t \notin v^-(q)$, as well as $t \in v^+(p)$. Clearly, $\mathcal{M}, t \vDash^+ p$, but $\mathcal{M}, t \not\vDash^+ q \vee \sim q$, establishing $p \not\vDash_C q \vee \sim q$ hence the paracompleteness of C.

□

3.2.6 Classical validities invalidated by C

Proposition 3.2.8 (invalidating contraposition).

$$\not\vDash_C (\varphi \to \psi) \to (\sim \psi \to \sim \varphi)$$

Proof. The following model is a counter-model to the instance $(p \to q) \to (\sim q \to \sim p)$ of contra-position, for p, q distinct atomic formulas.

Let $W = \{t, s\}$ with $t \leq s$. Let

$$\{t, s\} \subseteq v^+[\![p]\!] \quad t \in v^-[\![p]\!] \quad s \notin v^-[\![p]\!]$$

$$\{t, s\} \subseteq v^+[\![q]\!] \quad \{t, s\} \subseteq v^-[\![q]\!]$$

Therefore, $\mathcal{M} \vDash_C p \to q$, but $\mathcal{M} \not\vDash_C \sim q \to \sim p$.

□

3.2.7 Variations on C

In this section, I review some of the extensions of C, that shed some more light on connexivity.

3.2.7.1 *MC*

In [122], Wansing presents a variant MC of C, *material Connexive logic*, The logic MC has similar properties to those of C, but does not rely on strong negation, producing a "material-like" connexive conditional. It also is negation inconsistent but paraconsistent.

Models for MC are classical-like (without strong negation), and connexivity is again obtained by modifying the falsification condition for the conditional.

Satisfaction of φ in a model $\mathcal{M}$ is defined as follows, with v a valuation of $\{1, 0\}$ to atomic formulas.

$$\begin{aligned}
\mathcal{M} &\models p & &\text{iff} & v[\![p]\!] &= 1 \\
\mathcal{M} &\models \varphi \wedge \psi & &\text{iff} & \mathcal{M} &\models \varphi \text{ and } \mathcal{M} \models \psi \\
\mathcal{M} &\models \varphi \vee \psi & &\text{iff} & \mathcal{M} &\models \varphi \text{ or } \mathcal{M} \models \psi \\
\mathcal{M} &\models \varphi \to \psi & &\text{iff} & \mathcal{M} &\not\models \varphi \text{ or } \mathcal{M} \models \psi \\
\\
\mathcal{M} &\models \neg p & &\text{iff} & v[\![\neg p]\!] &= 1 \\
\mathcal{M} &\models \neg\neg\varphi & &\text{iff} & \mathcal{M} &\models \varphi \\
\mathcal{M} &\models \neg(\varphi \wedge \psi) & &\text{iff} & \mathcal{M} &\models \neg\varphi \text{ or } \mathcal{M} \models\sim \psi \\
\mathcal{M} &\models \neg(\varphi \vee \psi) & &\text{iff} & \mathcal{M} &\models \neg\varphi \text{ and } \mathcal{M} \models\sim \psi \\
\mathcal{M} &\models \neg(\varphi \to \psi) & &\text{iff} & \mathcal{M} &\not\models \varphi \text{ or } \mathcal{M} \models \neg\psi
\end{aligned} \qquad (3.2.9)$$

Validity is defined as satisfaction in every model, and similarly for logical consequence.

3.2 Wansing's propositional Connexive logic C

Alternatively, MC can be defined by the following four-valued truth-tables over $\mathcal{V} = \{t, b, n, f\}$, known from FDE (see Section 7.4.1.2).

φ	$\neg\varphi$
t	f
b	b
n	n
f	t

$\varphi \wedge \psi$	t	b	n	f
t	t	b	n	f
b	b	b	f	f
n	n	f	n	f
f	f	f	f	f

$\varphi \vee \psi$	t	b	n	f
t	t	t	t	t
b	t	b	t	b
n	t	t	n	n
f	t	b	n	f

$\varphi \rightarrow \psi$	t	b	n	f
t	t	b	n	f
b	t	b	n	f
n	b	b	b	b
f	b	b	b	b

(3.2.10)

Finally, axiomatically MC is defined like C (cf. (3.2.2)), but extending propositional Classical logic instead of propositional Intuitionistic logic.

The logic MC is further investigated by Omori and Wansing [88] by considering various effects of extending it to a Connexive logic $C3$ by "closing the gap", i.e., removing the truth-value b (both true and false), and adding (*LEM*), $\varphi \vee \neg\varphi$, in connection with Cantwell's logic CN.

3.2.7.2 2C

In [121], Wansing presents a connexive extension of C, $2C$, that introduces a twist into the plot: the logic has an additional conditional-like connective, *co-implication*. The connexive version of co-implication in this logic is *dual* to the connexive implication.

I consider this logic in some more detail in Section 8.2.2.

3.3 Priest's propositional Connexive logics

I now turn to the propositional Connexive logics introduced in by Priest [96]. Priest considers two versions of this logic, and I will use Fergusson's name in [29] for the one I prefer, namely, P_S (the "symmetrized" logic) which is both paraconsistent and paracomplete. The first-order extension of Priests Connexive logics (two variants thereof) are considered in Section 6.6.1.

The central idea driving the design of P_S is the view expressed by the slogan *'negation as cancellation'*, whereby $\neg\varphi$ "cancels" (or "retracts") the content of φ. Read this way, a contradiction $\varphi \wedge \neg\varphi$ has no contents at all, as the second conjunct cancels the content of the first conjunct. For a criticism of negation as cancelation as a basis for a Connexive logic see Wansing and Skurt [123].

In contrast to Wansing's method of obtaining connexivity by modifying the falsification conditions of the conditional, Priest obtains full (not humble only) connexivity by:

- Giving an account of negation as cancellation.
- Modifying the verification conditions of the conditional, and of the logical consequence relation the conditional internalizes.

3.3 Priest's propositional Connexive logics

The object-language is again like that for C, with the connectives '$\neg$' (negation), '$\wedge$' (conjunction), '$\vee$' (disjunction) and '$\rightarrow$' (conditional). The presentation is model-theoretic.

3.3.1 Models for P_S

A model for P_S is a structure $\mathcal{M} = \langle W, g, v \rangle$, where:

- W is a set of *worlds* (just another name for points of evaluation), ranged over by w. No ordering is assumed over W.

- $g \in W$ is a distinguished world.

- v is a family of assignment functions assigning a truth-value $v_w[\![p]\!] \in \{0, 1\}$ to every atomic proposition p in a world w.

The satisfaction of a formula φ in a model $\mathcal{M}$ at a world w, denoted $\mathcal{M}, w \vDash \varphi$, is defined recursively as follows.

$\mathcal{M}, w \vDash p$ iff $v_w[\![p]\!] = 1$

$\mathcal{M}, w \vDash \neg \varphi$ iff $\mathcal{M}, w \nvDash \varphi$

$\mathcal{M}, w \vDash \varphi \wedge \psi$ iff $\mathcal{M}, w \vDash \varphi$ and $\mathcal{M}, w \vDash \psi$

$\mathcal{M}, w \vDash \varphi \vee \psi$ iff $\mathcal{M}, w \vDash \varphi$ or $\mathcal{M}, w \vDash \psi$

$\mathcal{M}, w \vDash \varphi \rightarrow \psi$ iff $\exists w' \in W : \mathcal{M}, w' \vDash \varphi$ and $\exists w' \in W : \mathcal{M}, w' \vDash \neg \psi$ and $\forall w' \in W :$ if $\mathcal{M}, w' \vDash \varphi$ then $\mathcal{M}, w' \vDash \psi$
(3.3.11)

Thus, the vacuity of the conditional is eliminated! If φ is false in every world. then φ does not imply anything. Similarly, if ψ is true in every world, φ is not implied by anything. We see the null-paraconsistency and null-paracompleteness.

Such a definition of a conditional appears also in Gherardi and Orlandelli [40] as the *super-strict conditional*.

As for the biconditional, we get that $\mathcal{M}, w \models \varphi \leftrightarrow \psi$ requires that both of φ and ψ *and their negations* $\neg\varphi$ and $\neg\psi$ are each satisfied in some world, and then every world satisfies φ iff it satisfies ψ.

Validity is defined as satisfaction in the distinguished world g.

$$\mathcal{M} \models \varphi \text{ iff } \mathcal{M}, g \models \varphi \tag{3.3.12}$$

The definition is naturally extended point-wise to $\mathcal{M} \models \Gamma$ for Γ a collection of formulas.

As for logical consequence:

$$\Gamma \models_{P_S} \varphi \text{ iff } \begin{array}{l} \exists \mathcal{M} : \mathcal{M} \models \Gamma \text{ and} \\ \exists \mathcal{M} : \mathcal{M} \models \neg\varphi \text{ and} \\ \forall \mathcal{M} : \text{ if } \mathcal{M} \models \Gamma \text{ then } \mathcal{M} \models \varphi \end{array} \tag{3.3.13}$$

Indeed, null-paraconsistency and null-paracompleteness!

Remarks:

1. Negation is defined as in Classical logic, not as the strong negation in C.

2. The definition of negation clearly validates double-negation equivalence: $\mathcal{M}, w \models \varphi$ iff $\mathcal{M}, w \models \neg\neg\varphi$.

3. The connexive conditional '$\rightarrow$' is intensional here too, its satisfaction in a world w depending on all worlds (points) in W.

4. Clearly, contraposition holds: $\mathcal{M}, w \models \varphi \rightarrow \psi$ iff $\mathcal{M}, w \models \neg\psi \rightarrow \neg\varphi$.

3.3 Priest's propositional Connexive logics

5. No separate falsification condition is given for the conditional.

6. The logical consequence relation is both null-paraconsistent and null-paracomplete, reflecting the verification condition for the conditional.

7. The logical consequence is *non-monotonic*, as the addition of an assumption may render Γ inconsistent; furthermore, it is *non-reflexive*.

Because adding assumptions may render the collection of assumptions inconsistent, we get immediately the following corollary.

Corollary 3.3.6 (non-monotonicity of $\vDash_{P_S}$). $\Gamma \vDash_{P_S} \varphi$ *does not imply* $\Gamma, \Delta \vDash_{P_S} \varphi$.

Example 3.3.6 (non-monotonicity of $\vDash_{P_S}$ (Priest)). *It is easily verifiable that* $\varphi, \neg\varphi \vee \psi \vDash_{P_S} \psi$,
but $\varphi, \neg\varphi, \neg\varphi \vee \psi \nvDash_{P_S} \psi$, *since adding the assumption* $\neg\varphi$ *renders the assumptions inconsistent, not implying anything by the null account.* □

Another corollary of the modified logical consequence definition is the following.

Corollary 3.3.6 (non-closure under uniform substitution). $\vDash_{P_S}$ *is not closed under uniform substitution.*

Example 3.3.7 (non-closure under uniform substitution (Priest)). *Clearly* $p \vDash_{P_S} p$, *but* $\varphi \wedge \neg\varphi \nvDash_{P_S} \varphi \wedge \neg\varphi$. *The substitution generated a self-contradictory assumption, implying nothing, not even itself.* □

3.3.2 Connexivity of P_S

To see the connexivity of P_S, consider the following argument by Priest showing that P_S validates A_1 in (2.2.1) and $B_1^{\vDash}$.

Proposition 3.3.9 (connexivity of P_S). *For every φ and ψ:*

1. $\vDash_{P_S} \neg(\varphi \to \neg\varphi)$

2. $\varphi \to \psi \vDash_{P_S} \neg(\varphi \to \neg\psi)$

Proof. 1. Assume, towards a contradiction, that for some $\mathcal{M}$, it holds that $\mathcal{M} \vDash \varphi \to \neg\varphi$. Thus, the following two conditions should hold by the definition of satisfaction: (1) for some $w \in W$, $\mathcal{M}, w \vDash \varphi$ and (2) $\forall w \in W$: if $\mathcal{M}, w \vDash \varphi$ then $\mathcal{M}, w \vDash \neg\varphi$, which is impossible.

2. First, it is easy to observe that both $\varphi \to \psi$ and $\varphi \to \neg\psi$ (the negation of the conclusion) are satisfiable.

Let $\mathcal{M}$ be a model of $\varphi \to \psi$. Suppose, towards a contradiction, that $\mathcal{M} \vDash \varphi \to \neg\psi$. Then, for every $w \in W$, both if $\mathcal{M}, w \vDash \varphi$ then $\mathcal{M}, w \vDash \psi$ and $\mathcal{M}, w \vDash \varphi$ then $\mathcal{M}, w \vDash \neg\psi$; this is impossible since there is a $w \in W$ s. t. $\mathcal{M}, w \vDash \varphi$.

□

The other characteristics of connexivity are established similarly.

Thus, what brings about the connexivity of P_S is the adoption of the null account in the definition of logical consequence and, in parallel, the truth condition for the conditional.

3.3.3 Paraconsistency and paracompleteness of P_S

As already mentioned above, the modified definition of logical consequence (the null account) renders P_S both paraconsistent and paracomplete.

3.4 Angell's propositional Connexive logic

Proposition 3.3.10 (paraconsistency and paracompleteness of P_S).

$$(1)\ \varphi \wedge \neg \varphi \not\vDash_{P_S} \psi \quad (2)\ \varphi \not\vDash_{P_S} \psi \vee \neg \psi \tag{3.3.14}$$

3.4 Angell's propositional Connexive logic

3.4.1 Introduction

In [2], Angell presents a Connexive logic $\mathbf{PA_1}$, explicitly intended to capture subjunctive conditionals in NL. The B_i axioms are seen as *principles of subjunctive contrariety*. The logic preserves also the material conditional '⊃', viewed as an abbreviation of one of its classical equivalents, and introduces in addition another conditional '→', a connexive conditional, which avoids the paradoxes of the material conditional.

The negation consistency of the axioms of $\mathbf{PA_1}$ is established by presenting 4-valued truth-tables for the connectives (Section 3.4.3), which validate all the axioms of $\mathbf{PA_1}$.

3.4.2 The axiomatic presentation of PA_1

The axioms of PA_1 are presented in Figure 3.1, notationally adjusted.

Remarks:

1. Note that the axiom B_1 is an axiom of PA_1, namely $A10$. It is the only non-classically valid axiom in the system.

2. Many of the "non-disturbing" classical theses are provable in $\mathbf{PA_1}$.

A1. $((\psi \to \chi) \to ((\varphi \to \psi) \to (\varphi \to \chi)))$

A2. $(\varphi \to \psi) \to ((\chi \wedge \varphi) \to (\psi \wedge \chi)))$

A3. $((\varphi \to \neg(\psi \wedge \chi)) \to ((\psi \wedge \varphi) \to \neg \chi))$

A4. $(\varphi \wedge \psi) \wedge \chi \to (\varphi \wedge (\psi \wedge \chi))$

A5. $(\varphi \to \neg \psi) \to (\psi \to \neg \varphi)$

A6. $\neg \neg \varphi \to \varphi$

A7. $(\varphi \to \psi) \to \neg(\varphi \wedge \neg \psi)$

A8. $\neg((\varphi \wedge \psi) \wedge \neg \varphi)$

A9. $\neg(\varphi \wedge \neg(\varphi \wedge \varphi))$

A10. $(\varphi \to \psi) \to \neg(\varphi \to \neg \psi)$

$$\dfrac{\vdash \varphi \to \psi \quad \vdash \varphi}{\vdash \psi} \, (MP) \quad \dfrac{\vdash \varphi \quad \vdash \psi}{\vdash \varphi \wedge \psi} \, (ADJ) \quad \dfrac{\vdash \varphi}{\vdash \varphi(\chi/p))} \, (USUB), \, p \in At(\varphi)$$
(3.4.15)

Figure 3.1: The axioms of **PA**$_1$

3.4 Angell's propositional Connexive logic

Example 3.4.8 (identity).

$$\vdash_{\mathbf{PA}_1} \varphi \to \varphi \qquad (3.4.16)$$

The derivation is:

1. $((\psi \to \chi) \to ((\varphi \to \psi) \to (\varphi \to \chi)))$ $(A1)$
2. $((\neg\neg\varphi \to \varphi) \to ((\varphi \to \neg\neg\varphi) \to (\varphi \to \varphi)))$ $(1., [\neg\neg\varphi/\psi, \varphi\chi])$
3. $\neg\neg\varphi \to \varphi$ $(A6.)$
4. $(\varphi \to \neg\neg\varphi) \to (\varphi \to \varphi)$ $[1., 3., MP]$
5. $\varphi \to \neg\neg\varphi$ $[A previous thesis]$
6. $\varphi \to \varphi$ $[4., 5., MP]$

I use instantiation of meta-variables where originally the uniform substitution rule (USUB) is used.

3.4.2.1 Derivability of the connexive axioms

Next, to establish connexivity of PA_1, I show the derivation of the connexive axioms A_1 and A_2.

Proposition 3.4.11 (A1). $\vdash_{\mathbf{PA}_1} \neg(\varphi \to \neg\varphi)$

Proof. The derivation is:

1. $(\varphi \to \psi) \to \neg(\varphi \to \neg\psi)$ $[A10.]$
2. $(\varphi \to \varphi) \to \neg(\varphi \to \neg\varphi)$ $[1., \varphi/\psi]$
3. $\varphi \to \varphi$ $[thesis\ proved\ above]$
4. $\neg(\varphi \to \neg\psi)$ $[2., 3., MP]$

□

Proposition 3.4.12 (A2). $\vdash_{\mathbf{PA}_1} \neg(\neg\varphi \to \varphi)$

φ	$\neg\varphi$
3	0
2	1
1	2
0	3

$\wedge$	0	1	2	3
0	1	0	3	2
1	0	1	2	3
2	3	2	3	2
3	2	3	2	3

$\rightarrow$	0	1	2	3
0	1	2	3	2
1	2	1	2	3
2	1	2	1	2
3	2	1	2	1

(3.4.17)

Figure 3.2: The four-valued truth-tables for $\mathbf{PA}_1$

Proof. The derivation is

1. $(\varphi \rightarrow \neg\psi) \rightarrow (\psi \rightarrow \neg\varphi)$ [A5]
2. $((\varphi \rightarrow \psi) \rightarrow \neg(\varphi \rightarrow \neg\psi)) \rightarrow ((\varphi \rightarrow \neg\psi) \rightarrow \neg(\varphi \rightarrow \psi))$ [1., $\varphi \rightarrow \psi/\varphi$, $\varphi \rightarrow \neg\psi/\psi$]
3. $(\varphi \rightarrow \psi) \rightarrow \neg(\varphi \rightarrow \neg\psi)$ [A10]
4. $((\varphi \rightarrow \neg\psi) \rightarrow \neg(\varphi \rightarrow \psi))$ [2., 3., MP]
5. $(\neg\varphi \rightarrow \psi) \rightarrow (\neg(\neg\varphi \rightarrow \neq \psi)$ [3., $\neg\varphi/\varphi$]
6. $(\neg\varphi \rightarrow \neg\psi) \rightarrow (\neg(\neg\varphi \rightarrow \psi)$ [4., $\neg\varphi/\varphi$]
7. $(\neg\varphi \rightarrow \neg\varphi) \rightarrow (\neg(\neg\varphi \rightarrow \varphi)$ [6., φ/ψ]
8. $\neg\varphi \rightarrow \neg\varphi$ [(3.4.16) $\neg\varphi/\varphi$]
9. $\neg(\neg\varphi \rightarrow \varphi)$ [7., 8., MP]

□

There is also a $\mathbf{PA}_1$-derivation of B_2 in Angell [2] (p. 335).

3.4.3 4-valued truth-tables for $\mathbf{PA}_1$

The model-theory of $\mathbf{PA}_1$ is given by means of truth-tables for the connectives over four truth-values, $\mathcal{V} = \{3, 2, 1, 0\}$, as shown in Figure 3.2. The *designated truth-values* are $\mathcal{D} = \{0, 1\}$. Angell himself does not provide an interpretation of the four truth-values he uses. Ferguson

3.4 Angell's propositional Connexive logic

[30] quotes Routley and Montgomery, giving an interpretation that they admit is problematic:

> $CC1$, for instance, can be given a semantics by associating the matrix value [0] with logical necessity, value [3] with logical impossibility, value [1] with contingent truth, and value [2] with contingent falsehood.

Notably, a conditional never gets the value 0.

Definition 3.4.2 (logical consequence). *for every Γ and φ:*

1. *φ is a $\mathbf{PA_1}$-consequence of Γ, denoted $\Gamma \vDash_{\mathbf{PA_1}} \varphi$ iff for any valuation $\mathbf{v}$ s.t. $\mathbf{v}[\![\psi]\!] \in \mathcal{D}$ for every $\psi \in \Gamma$, also $\mathbf{v}[\![\varphi]\!] \in \mathcal{D}$.*

2. *Consequently, φ is $\mathbf{PA_1}$-valid, denoted $\vDash_{\mathbf{PA_1}} \varphi$. iff for every valuation $\mathbf{v}$, $\mathbf{v}[\![\varphi]\!] \in \mathcal{D}$.*

The soundness and completeness of the $\mathbf{PA_1}$ axioms and rules are shown (somewhat indirectly) in Angell [2]. I skip the details here.

3.4.3.1 Non-derivability of some Classical logic validities

Proposition 3.4.13 (non-derivability of the paradoxes of material conditional). *For every φ, ψ:*

1.
$$\nvDash_{\mathbf{PA_1}} \varphi \to (\psi \to \varphi) \tag{3.4.18}$$

2.
$$\nvDash_{\mathbf{PA_1}} \neg \varphi \to (\varphi \to \psi) \tag{3.4.19}$$

Proof. The non-$\mathbf{PA}_1$-derivability follows from the non-$\mathbf{PA}_1$-validity by soundness.

1. Let v be s.t. $\mathbf{v}[\![\varphi]\!] = 0$ and $\mathbf{v}[\![\psi]\!] = 2$. Then, $\mathbf{v}[\![\varphi \to (\psi \to \varphi)]\!] = 2 \notin \mathcal{D}$, showing
$$\nvdash_{\mathbf{PA}_1} \varphi \to (\psi \to \varphi)$$

2. Let v be such that $\mathbf{v}[\![\varphi]\!] = 0$ and $\mathbf{v}[\![\psi]\!] = 1$. Then, $\mathbf{v}[\![\neg\varphi \to (\varphi \to \psi)]\!] = 2 \notin \mathcal{D}$, showing
$$\nvdash_{\mathbf{PA}_1} \neg\varphi \to (\varphi \to \psi)$$

□

Proposition 3.4.14 (non-derivability of conjunction simplification). *For every φ, ψ:*

1.
$$\nvdash_{\mathbf{PA}_1} \varphi \wedge \psi \to \varphi \qquad (3.4.20)$$

2.
$$\nvdash_{\mathbf{PA}_1} \varphi \wedge \psi \to \psi \qquad (3.4.21)$$

Proof. 1. Let v be s.t. $\mathbf{v}[\![\varphi]\!] = 1$ and $\mathbf{v}[\![\psi]\!] = 0$. Then, $\mathbf{v}[\![\varphi \wedge \psi \to \varphi]\!] = 2 \notin \mathcal{D}$, showing
$$\nvdash_{\mathbf{PA}_1} \varphi \wedge \psi \to \varphi$$

2. Let v be s.t. $\mathbf{v}[\![\varphi]\!] = 0$ and $\mathbf{v}[\![\psi]\!] = 1$. Then, $\mathbf{v}[\![\varphi \wedge \psi \to \varphi]\!] = 2 \notin \mathcal{D}$, showing
$$\nvdash_{\mathbf{PA}_1} \varphi \wedge \psi \to \psi$$

□

Proposition 3.4.15 (non-derivability of symmetric implication).
$$\nvdash_{\mathbf{PA}_1} (\varphi \to \psi) \to (\psi \to \varphi)$$

3.4 Angell's propositional Connexive logic

Proof. Let p, q be two distinct atomic propositions. Let **v** be s.t. $\mathbf{v}[\![p]\!] = 2$ and $\mathbf{v}[\![q]\!] = 0$. Then, $\mathbf{v}[\![p \to q]\!] = 1$ and $\mathbf{v}[\![q \to p]\!] = 3$, resulting in $\mathbf{v}[\![(p \to q) \to (q \to p)]\!] = 3 \notin \mathcal{D}$.
Hence, $\not\models (p \to q) \to (q \to p)$. □

Remark 3.4.7 (failure of disjunction addition). *If disjunction is defined by*
$$\varphi \lor \psi \equiv \neg(\neg\varphi \land \neg\psi)$$
disjunction addition fails too.

□

Remark 3.4.8. *In a later paper [3], Angell notes that PA_1, as well as McCall's $CC1$ (see Section 3.6.2) contain the following undesired thesis:*
$$\vdash_{PA_1} (\varphi \to \psi) \land (\chi \to \psi) \to \neg(\varphi \to \neg\chi)$$
It validates the following intuitively invalid argument.

> if x is a dog then x is an animal
> if x is a cat then x is an animal
> Therefore ¬(if x is a dog then x is not a cat)

As a (partial) remedy, Angell produced a variant PA_2 with different truth-tables.

□

Proposition 3.4.16 (negation consistency of PA_1). *PA_1 is negation consistent.*

Proof. Assume $\vdash_{PA_1} \varphi$. Consider any PA_1-valuation **v**. By assumption, $\mathbf{v}[\![\varphi]\!] \in \mathcal{D} = \{0, 1\}$. By the truth-table of negation, $\mathbf{v}[\![\neg\varphi]\!] \in \{2, 3\}$; that is, $\mathbf{v}[\![\neg\varphi]\!] \notin \mathcal{D}$. Hence, $\not\models_{PA_1} \neg\varphi$, and by the soundness of PA_1 $\not\vdash_{PA_1} \neg\varphi$. □

Consequently, PA_1 validates explosion.

3.5 Francez's Connexive logics

3.5.1 Introduction

In this section, I present two propositional Connexive logics defined by two natural-deduction proof-systems $\mathcal{N}^{\neg r}$ and $\mathcal{N}^{\neg l}$, originally presented in Francez [33]. This definitional tool differs from the axiomatic or model-theoretic definitions employed for defining the previously encountered Connexive logics.

These Connexive logics are inspired by a certain use of negation and conditional in natural language, leading to splitting the negation into two different negations, $\neg_r$ and $\neg_l$. A model-theory (in Section 3.5.7) is used mainly as an auxiliary tool for establishing non-derivability, for example of some classical formal theorems (or, more generally, classical derivability claims) that are not provable (not derivable) in $\mathcal{N}^{\neg r}$ and $\mathcal{N}^{\neg l}$. The relationship between the conditional and the negation in the system $\mathcal{N}^{\neg r}$ is similar to the one in Cantwell [13] CN and Cooper [14] OL, the former unaware of the latter.

Having *all* the B_i axioms and their converses live together, and having a transitive conditional, leads to certain undesired complications, related to introducing and eliminating the operators. Identifying the two negations and creating one negation having the properties of both would:

- blur the distinction between the views of implication as focusing on sufficiency in contrast to focussing on necessity, a driving force behind the proposed systems, as discussed below. In particular,[3] $\varphi \rightarrow \psi$ becomes both necessary and sufficient for $\neg\varphi \rightarrow \neg\psi$,

---
[3] I thank Heinrich Wansing for this observation.

3.5 Francez's Connexive logics

as shown by the following derivations.

$$\cfrac{\cfrac{\cfrac{}{(\varphi\to\psi)\to\neg(\neg\varphi\to\psi)}\,(B_3)}{\neg(\neg\varphi\to\psi)\to(\neg\varphi\to\neg\psi)}\,(conv.\ B_2)}{(\varphi\to\psi)\to(\neg\varphi\to\neg\psi)}\,(Trans\to)$$

$$\cfrac{\cfrac{\cfrac{}{(\neg\varphi\to\neg\psi)\to\neg(\varphi\to\neg\psi)}\,(B_4)}{\neg(\varphi\to\neg\psi)\to(\varphi\to\psi)}\,(conv.\ B_1)}{(\neg\varphi\to\neg\psi)\to(\varphi\to\psi)}\,(Trans\to)$$

This equivalence does not conform to the standard meanings of sufficiency and necessity.

- render negation *ambiguous*, certainly an undesired effect.

Therefore, I "split" the negation into two[4] negations, '$\neg_l$' and '$\neg_r$', each separately responsible to one of the A_i connexive axioms and two of the B_is, reformulated as formal theorems in terms of the two negations.

However, see Section 4.3 for another approach, allowing to keep just one negation.

Below, all the propositional connexive axioms are shown as formal theorems either of the ND-system $\mathcal{N}^{\neg r}$ or of $\mathcal{N}^{\neg l}$.

When viewed from a model-theoretic perspective, neither of those negations here is a *contradiction-forming* operator, except when applied to atomic propositions (justified below in Section 3.5.4). Rather, both are sub-contrariety formers (in two different ways). They play two separate roles for compound propositions of the form $\alpha\to\beta$, distinguished as described below. Recall that a generic conditional $\alpha\to\beta$ can be read in two ways.

---

[4] Note that while I use the same technical term, split negation here is unrelated to the split negation in e.g., Sequoiah-Grayson [111], the latter resulting from non-commutativity of the logic.

- α is sufficient for β.

- β is necessary for α.

The two negations negate those two readings in the way described below.

- From the NL point of view, '$\neg_r$' is a "corrective negation", expressing disagreement about *sufficiency* of the condition α, taking it as being sufficient for $\neg\beta$ instead of being sufficient for β.

- From the NL point of view, '$\neg_l$' too is a "corrective negation", expressing disagreement about *necessity* of the condition β for α, taking instead β as being necessary for $\neg\alpha$.

Thus, the A_is express the impossibility of φ to be either necessary or sufficient for its own negation. If necessity and sufficiency are endowed a non-truth-functional meaning, one based on contents, then this interpretation of the conditional and negation expresses relationships between sentential meanings transcending the simple classical truth-functionality of the conditional and the negation.

Throughout this section, I consider a propositional fragment containing, in addition to atomic propositions, conditional and negation only.

3.5.2 Natural language motivation

3.5.2.1 Negating conditionals

The point of departure is the following schematic dialog D between two participants A and B using two *compound* formulas (i.e., non-atomic,

3.5 Francez's Connexive logics

headed by a generic conditional '$\to$') α and β in the following way:

$$D :: \begin{array}{l} A:\ \alpha \\ B:\ \text{No!}\ \beta \end{array} \qquad (3.5.22)$$

Here α is $\varphi \to \psi$ (for some φ, ψ), while β is either $\neg\varphi \to \psi$ or $\varphi \to \neg\psi$. At this stage '$\neg$' is also considered a generic negation, to be made specific below.

The intended reading of the dialog D is characterized by the following two characteristics:

1. Participant B, by using No, *partially disagrees* with A about α by "negating" the latter.

2. Participant B offers β as the negated α expressing[5] the disagreement and "correcting" it.

Note that the "corrections" express consent about one argument of '$\to$', while negating the other argument of '$\to$'. I will refer to the arguments of '$\to$'s the left and right arguments, which explains the labels of the two negations. Clearly, this way of negating, *not* by contradicting, excludes the intuitionistic way of defining negation as implying $\bot$, absurdity. It fits the view of negating the conditional as expressed in Section 2.4.

I also consider a kind of a dual dialogue, in which one of the arguments of '$\to$' is already negated. That is, α is $\neg\varphi \to \psi$ or $\varphi \to \neg\psi$ (for some φ, ψ), while β is, respectively, $\neg(\varphi \to \psi)$. This suggest that double-negation elimination is implicitly employed in the "correction".

Throughout this section, '$\vdash$' refers to derivability in the natural-deduction system $\mathcal{N}^{\neg r}$ and $\mathcal{N}^{\neg l}$, to be presented below, and '$\dashv\vdash$' refers to mutual derivability. The context determines which proof-system is intended.

---

[5]In a naturally occurring dialog of the type D, a certain focal stress might be required. I ignore here such matters.

I will first consider the negation of compound formulas, deferring to Section 3.5.4 the definition of negating an atomic proposition, using just $\neg p$ (unsubscripted) for its expression.

The ND-system $\mathcal{N}^{\neg r}$ and $\mathcal{N}^{\neg l}$ below induce the following mutual derivability, a variant on hyper-connexivity, which can be interpreted as manifesting the sub-contrariness formation by the two negations.

$$\neg_l(\varphi \to \psi) \dashv\vdash \neg_l\varphi \to \psi \qquad \neg_r(\varphi \to \psi) \dashv\vdash \varphi \to \neg_r\psi \qquad (3.5.23)$$

The inspiration from the natural language dialog D pertains more to the first-degree case (without nesting of conditionals), but the incorporation of the generalization with unrestricted nesting into the object-language is needed in order to obtain a logic.

Before turning to a general theory, I will consider in some detail some instances of the dialog D, to get a better intuition about what is involved in a partial disagreements of the intended type. In Section 3.5.3 I correlate the D-dialogs to Ramsey's test and a *dual* of this test.

Example 3.5.9. *I return to Cantwell's Example 2.4.4. Cantwell's motivation is completely different. His intention is to remove the feature of the material conditional of yielding a truth-value (actually, yielding the value 'true') in case the antecedent of the conditional yields a truth-value 'false'. The central feature of the interaction between negation and the conditional is the satisfaction of the following relation*

$$\neg(\varphi \to \psi) \vdash \varphi \to \neg\psi \qquad (3.5.24)$$

This reflects a common view that the truth of an antecedent of a conditional is a *presupposition* of asserting that conditional. It remains a presupposition also of the assertion of the negated conditional.

I will strengthen this relation, in accord with (3.5.23), into the hyper-connexive:

$$\neg(\varphi \to \psi) \dashv\vdash \varphi \to \neg\psi \qquad (3.5.25)$$

3.5 Francez's Connexive logics

Such a relationship, with a biconditional instead of mutual derivability, is essential to the modal Connexive logic C introduced by Wansing (see [119, p. 371]), as described in Section 3.2.

The way the connectives are defined by Cantwell and made to satisfy (3.5.24) is via a model theory based on a certain three-valued logic. Without being aware of, he uses *the same* three-valued truth-tables for the conditional and the negation as does Cooper [14] for his logic OL.

Cantwell's exemplary dialogue featuring this interaction in Example 2.4.4 (not structured as D here) is repeated here for convenience.
Anne: **If Oswald didn't kill Kennedy, Jack Ruby did.**
Bill: **No! You're wrong.**
Recall what Cantwell says ([13, p. 246]) about this exchange:

> When Bill denies the conditional asserted by Anne, he neither asserts nor denies that Oswald did the killing (he can continue, "**If Oswald didn't kill Kennedy, Castro did**"); his denial seemingly amounts to no more than the assertion that if Oswald didn't shoot Kennedy then neither did Jack Ruby. This kind of "conditional denial" seems to be a basic move in the language game; conditional negation is the sentential operator that corresponds to this form of conditional denial: "**It is not the case that if Oswald didn't shoot Kennedy, Jack Ruby did.**"

Not mentioned by Cantwell, this interaction between a conditional and a negation is one of the characteristics of Connexive logics as discussed in Section 2.4.

As I stated above, I want to approach the whole topic proof-theoretically, with no reference to truth-values, neither two nor any other number of them, or to a relational frame model-theory.

The examples below all use negated atomic propositions only, their negation understood informally by now (to be presented in more detail in Section 3.5.4).

Example 3.5.10. *Consider a dialogue structured as D, featuring a negated conditional. Suppose participants A and B are fans of the same soccer team T, but have opposing opinions as to how well T is prepared to play in a bad weather.*

$$D_1 :: \quad \begin{array}{l} A: \text{If it rains, } T \text{ will win} \\ B: \text{No! If it rains, } T \text{ will not win} \end{array} \quad (3.5.26)$$

That is, B consents about it raining, but disagrees as to what is raining a sufficient condition for. Considering (3.5.24) as reflecting this instance of D is best presented as

$$\neg_r(p \to q) \dashv\vdash p \to \neg_r q \quad (3.5.27)$$

In the dual dialogue, we have

$$\hat{D}_1 :: \quad \begin{array}{l} B: \text{If it rains, } T \text{ will not win} \\ A: \text{No! If it rains, } T \text{ will win} \end{array} \quad (3.5.28)$$

represented as

$$\neg_r(p \to \neg_r q) \dashv\vdash p \to q \quad (3.5.29)$$

Here the effect of double negation elimination is manifested. □

Similar arguments, motivated by the way the conditional and the negation interact, are put forward in Cooper [14].

Example 3.5.11. *Consider another instance of D between the same participants, the fans of team T.*

$$D_2 :: \quad \begin{array}{l} A: \text{If it rains, } T \text{ will win} \\ B: \text{No! If it does not rain, } T \text{ will win} \end{array} \quad (3.5.30)$$

3.5 Francez's Connexive logics

Here B consents to team T winning, but disagrees about what the sufficient condition for that is. This can be modeled by[6]

$$\neg_l(p \to q) \dashv\vdash \neg_l p \to q \qquad (3.5.31)$$

Again, in the dual dialogue, we have

$$\hat{D}_2 :: \begin{array}{l} B: \text{If it does not rain, } T \text{ will win} \\ A: \text{No! If it rains, } T \text{ will win} \end{array} \qquad (3.5.32)$$

represented as

$$\neg_l(\neg_l p \to q) \dashv\vdash p \to q \qquad (3.5.33)$$

Here too is the effect of double negation elimination manifested.

In a naturally occurring instance of dialogs like $\mathcal{D}_2$, an intonational stress on 'does not' *will occur.*

Note that there is no claim that every instance of $\mathcal{D}_2$ is justified. For example,

$$D_3^* :: \begin{array}{l} A: \text{If Nissim is in Nahariya}^7, \text{ Nissim is in italy} \\ B: \text{No! If Nissim is not in Nahariya, Nissim is in italy} \end{array}$$
$$(3.5.34)$$

In $\mathcal{D}_3^*$, the corrective negation is not *justified. A more refined analysis of such dialogs should involve the* alternatives *to the sentences occurring in them, an analysis that would take me too much afar. All I care about here is that* there are *situations justifying the kind of negation as in (refeq:lneg).* □

In both the examples above, No expresses a way of negating a conditional different from the standard way of negating the material conditional.

---

[6]This kind of conditional is not considered by Cantwell.

3.5.3 Ramsey's test and a dual test

In [100, p. 155], Ramsey proposes the following argument (quoted below with a slight modification of notation to fit the current presentation) as an interpretation of negating the conditional in NL.

> If two people are arguing 'If φ will ψ?' and are both in doubt as to φ, they are adding φ hypothetically to their stock of knowledge and arguing on that basis about ψ; so that in a sense 'If φ, ψ' and 'If $\varphi, \neg\psi$' are contradictories.

The scenario described in the above paragraph fits the structure of the argument between participants A and B in dialog D_1 (cf. (3.5.26)) used to motivate '$\neg_r$'. It exactly reflects an argument as to what is φ sufficient for: ψ or $\neg\psi$.

The connection of negating the conditional to Ramsey's test was noted also by Ferguson [28].

I suggest a *dual* to Ramey's test with a scenario fitting the argument D_2 (cf. (3.5.30)), used to motivate '$\neg_l$', that reflects an argument between A and B as to for which of $\varphi, \neg\varphi$ is ψ necessary for.

A dual Ramsey test:

> If two people are arguing 'If φ will ψ?' and are both in doubt as to φ, they are adding $\neg\varphi$ hypothetically to their stock of knowledge and arguing on that basis about ψ; so that in a sense 'If φ, ψ' and 'If $\neg\varphi, \psi$' are contradictories.

This dual test exactly reflects an argument as to what is ψ necessary for: φ or $\neg\varphi$.

3.5.4 Negating atomic propositions

Since atomic propositions are not conditionals, considerations like distinguishing between focus on sufficient conditions and necessary conditions do not apply to them and cannot drive the definition of their negation.

Consider a dialogue $\hat{\mathcal{D}}$, structured similarly to $\mathcal{D}$ (cf. (3.5.22)), where α and β are both atomic propositions.

Example 3.5.12 (atomic dialogue:).

$$\hat{\mathcal{D}} :: \quad \begin{array}{l} A: T \text{ will win} \\ B: \text{No! } T \text{ will not win} \end{array}$$

Here participant B plainly disagrees with participant A's assessment about the outcome of a game involving team T, not involving any conditionality. Here B's correction of A's statement (following his NO!) is just a claim of the opposite proposition, clearly attempting to contradict A. □

This example exemplifies the idea behind defining

$$\neg_l p = \neg_r p = \neg p \qquad (3.5.35)$$

(where '$\neg$' is classical negation). This will lead to the ND-rules for atomic proposition in Section 3.5.5 to coincide with the classical ones.

3.5.5 The natural-deduction systems $\mathcal{N}^{\neg r}$ and $\mathcal{N}^{\neg l}$

The design of the ND-systems $\mathcal{N}^{\neg r}$ and $\mathcal{N}^{\neg l}$ in Figures 3.3 and 3.4, respectively, is based on the following principles.

1. Both negations, as mentioned above, are not contradiction-forming, except when applied to atomic propositions. Rather, both $\neg_r(\varphi \rightarrow \psi)$

and $\neg_l(\varphi\to\psi)$ are sub-contraries of $\varphi\to\psi$. This behavior is very similar to what is known in the semantics of natural language as 'neg raising' (see Horn and Wansing [46] for discussion and references).

2. The negations and the conditional are not *independent*, and have to be understood *together*. Technically, this means that the I/E-rules for the conditional are *not pure* (i.e., refer to more than one operator, see Dummett [19]). In the model-theory in Section 3.5.7, the dependence between the negations and the conditional results in a non-compositionality in the assignment of truth-value.

3. Negations here are *non-uniform*, their (I/E)-rules depending on the negated formula. There are no rules that might be seen as '$(\neg_r I)$' and '$(\neg_l I)$' by which $\neg_r\varphi$ and $\neg_l\varphi$ can be introduced for a "bare" compound φ. The negations '$\neg_r$' and '$\neg_l$' can only be introduced for a conditional, and this can be done in two ways. Accordingly, the two negated conditionals are eliminated differently. This is a major reason for the need for splitting the negation.

   One result of this non-uniformity of negation is that both the ND-systems introduced below do not admit the rule of *uniform substitution*. Atomic propositions are *not* propositional variables. This is also reflected in the definition of valuations in the model-theory (see Definition 3.5.3).

4. The two systems $\mathcal{N}^{\neg r}$ and $\mathcal{N}^{\neg l}$ cannot conveniently be amalgamated into one combined system. There is an issue of how to propagate negation to the appropriate argument of the conditional. Suppose one considers (in an alleged combined system) $\neg_r(\varphi\to\psi)$ by negating ψ. Which nested negation should be employed? In principle, both ways can do, leading both to $\varphi\to\neg_l\psi$ and $\varphi\to\neg_r\psi$. A similar situation pertains to $\neg_l(\varphi\to\psi)$, leading either to $\neg_r\varphi\to\psi$ or to $\neg_l\varphi\to\psi$. This does not lead to a coherent interpretation of the A_is and B_is, that should relate to *one and the same negation*

3.5 Francez's Connexive logics

each. This reinterpretation of the A_is and B_is is seen clearly in the separate systems.

5. While iterating the same negation makes sense, giving rise to two forms of double negation, both eliminable, the iterations

$$\neg_r \neg_l \varphi, \quad \neg_l \neg_r \varphi \tag{3.5.36}$$

do not seem to have an obvious interpretation. Those iterations are not well-formed if separation of systems is kept.

Additional remarks about the I/E-rules:

1. Classically, the double-negation I/E-rules are related to the reversing of truth-value associated with a contradiction-forming operator. Here, they originate from a different source. As negations are associated with disagreement about one of the arguments of '$\to$', when applied to an already negated argument, a negation finds, so to speak, nothing (that is, no conditional) to disagree about, so it cancels the disagreement when eliminated. When introduced, it can be seen, so to speak, as retracting the disagreement that would have been formed by negating once only.

2. Note that the '(dni)' rules are *primitive*, in contrast to classical '(dni)', which is derivable in Classical logic; this is again an effect of the current negations not being contradiction-forming, blocking the usual classical derivation of '(dni)'.

A methodological remark: In what sense can '$\neg_r$' and '$\neg_l$' "deserve" to be considered as negations? First, both are *involutive*, as is common for several other negations. Secondly, they are formed by a tool not used before in proof-theory: *negating a rule* (in contrast to the usual notion of negating a proposition). Here the rule classical/intuitionistic $(\to I)$, introducing an implication, is negated in two ways:

1. Negating the discharged assumption of the premise.

2. Negating the conclusion of the sub-derivation forming the premise.

As for ($\to E$), it is negated either by negating its minor premise or by negating its conclusion. Similar rules, but in a sequent calculus L/R-rules form, appear in Kamide and Wansing [51], but are not viewed as negating the standard L/R rules for a conditional.

Note that both $\mathcal{N}^{\neg r}$ and $\mathcal{N}^{\neg l}$ are *paraconsistent*, invalidating explosion (cf. Example 3.5.15). Neither one of $\varphi, \neg_r\varphi \vdash \psi$ and $\varphi, \neg_l\varphi \vdash \psi$ holds (except for an atomic φ). This is typical to sub-contrariety forming operators; see Béziau [10].

3.5.5.1 The ND-system $\mathcal{N}^{\neg r}$

Derivations (tree-shaped) are defined recursively as usual.

Proposition 3.5.17 (closure under composition). *Derivations in $\mathcal{N}^{\neg r}$ are closed under composition of derivations.*

The proof is standard, structured as in Negri and von Plato [76], and omitted.

Corollary 3.5.8. *The mutual derivability $\neg_r(\varphi \to \psi) \dashv\vdash_{\mathcal{N}^{\neg r}} \varphi \to \neg_r\psi$ (cf. (3.5.23)) holds.*

$\square$

Proof. The derivations are as follows.

$$\dfrac{\dfrac{\varphi \to \neg_r\psi \quad [\varphi]_1}{\neg_r\psi}(\to E)}{\neg_r(\varphi \to \psi)}(\neg_r I^1) \qquad \dfrac{\dfrac{\neg_r(\varphi \to \psi) \quad [\varphi]_1}{\neg_r\psi}(\neg_r \to E)}{\varphi \to \neg_r\psi}(\neg_r \to I^1) \qquad (3.5.41)$$

$\square$

3.5 Francez's Connexive logics

$$\frac{\begin{array}{c}[\varphi]_i\\ \vdots\\ \psi\end{array}}{\varphi\to\psi}\,(\to I^i) \qquad \frac{\varphi\to\psi \quad \varphi}{\psi}\,(\to E) \tag{3.5.37}$$

$$\frac{\begin{array}{c}[\varphi]_i\\ \vdots\\ \neg_r\psi\end{array}}{\neg_r(\varphi\to\psi)}\,(\neg_r\to I^i) \qquad \frac{\neg_r(\varphi\to\psi) \quad \varphi}{\neg_r\psi}\,(\neg_r\to E) \tag{3.5.38}$$

$$\frac{\begin{array}{cc}[p]_i & [p]_i\\ \vdots & \vdots\\ q & \neg_r q\end{array}}{\neg_r p}\,(At\neg_r I^i) \qquad \frac{p \quad \neg_r p}{\varphi}\,(At\neg_r E) \tag{3.5.39}$$

$$\frac{\neg_r\neg_r\varphi}{\varphi}\,(dne_r) \qquad \frac{\varphi}{\neg_r\neg_r\varphi}\,(dni_r) \tag{3.5.40}$$

Figure 3.3: The I/E-rules of $\mathcal{N}^{\neg_r}$

3.5.5.2 Derivability of the connexive axioms in $\mathcal{N}^{\neg r}$

I now show the the derivability of the connexive axioms in $\mathcal{N}^{\neg r}$, justifying its defining a Connexive logic.

Proposition 3.5.18 (Aristotele's $\neg_r$-thesis).

$$\vdash_{\mathcal{N}^{\neg r}} \neg_r(\varphi \to \neg_r \varphi) \tag{3.5.42}$$

$\square$

Proof. The $\mathcal{N}^{\neg r}$-derivation of is as follows.

$$\cfrac{\cfrac{[\varphi]_1}{\neg_r \neg_r \varphi}\ [(dni_r)]}{\neg_r(\varphi \to \neg_r \varphi)}\ (\neg_r {\to} I^1)$$

$\square$

Proposition 3.5.19 (Boethius' $\neg_r$-theses).

$$\begin{aligned}(B1)\ &\vdash_{\mathcal{N}^{\neg r}} (\varphi \to \psi) \to \neg_r(\varphi \to \neg_r \psi)\\ (B2)\ &\vdash_{\mathcal{N}^{\neg r}} (\varphi \to \neg_r \psi) \to \neg_r(\varphi \to \psi)\end{aligned} \tag{3.5.43}$$

$\square$

Proof. The $\mathcal{N}^{\neg r}$-derivation of (B1) is as follows.

$$\cfrac{\cfrac{\cfrac{\cfrac{[\varphi \to \psi]_2 \quad [\varphi]_1}{\psi}\ (\to E)}{\neg_r \neg_r \psi}\ (dni_r)}{\neg_r(\varphi \to \neg_r \psi)}\ (\neg_r {\to} I^1)}{(\varphi \to \psi) \to \neg_r(\varphi \to \neg_r \psi)}\ (\to I^2) \tag{3.5.44}$$

3.5 Francez's Connexive logics

The $\mathcal{N}^{\neg r}$-derivation of (B2) is as follows.

$$\dfrac{\dfrac{\dfrac{[\varphi \to \neg_r \psi]_2 \quad [\varphi]_1}{\neg_r \psi} (\to E)}{\dfrac{\neg_r(\varphi \to \psi)}{(\varphi \to \neg_r \psi) \to \neg_r(\varphi \to \psi)} (\to I^2)}}{} \quad (\neg_r \to I_2^1) \tag{3.5.45}$$

$\square$

Example 3.5.13.

$$\neg_r(\varphi \to (\psi \to \chi)) \vdash_{\mathcal{N}^{\neg r}} (\varphi \to (\psi \to \neg_r \chi))$$

The $\mathcal{N}^{\neg r}$-derivation is

$$\dfrac{\dfrac{\neg_r(\varphi \to (\psi \to \chi))}{(\varphi \to \neg_r(\psi \to \chi))} (3.5.23)}{(\varphi \to (\psi \to \neg_r \chi))} (3.5.23)$$

3.5.5.3 Negation inconsistency of $\mathcal{N}^{\neg r}$

Proposition 3.5.20 (Negation inconsistency of $\mathcal{N}^{\neg r}$)**.**

$$\vdash_{\mathcal{N}^{\neg r}} (((\neg_r \varphi \to \varphi) \to \neg_r \varphi) \to \neg_r((\neg_r \varphi \to \varphi) \to \neg \varphi)) \tag{3.5.46}$$

The result follows, since (3.5.46) is a contradiction to A_1.

Proof. The $\mathcal{N}^{\neg r}$-derivation is as follows.

$$\dfrac{\dfrac{\dfrac{\dfrac{[((\neg \varphi \to \varphi) \to \neg \varphi)]_1 \quad [\neg \varphi \to \varphi]_2}{\neg_r \varphi} (\to E) \quad [\neg \varphi \to \varphi]_2}{\dfrac{\varphi}{\neg_r \neg_r \varphi} (dni)} (\to E)}{\neg_r((\neg_r \varphi \to \varphi) \to \neg_r \varphi))} (\neg_r \to I^2)}{(((\neg_r \varphi \to \varphi) \to \neg_r \varphi) \to \neg_r((\neg_r \varphi \to \varphi) \to \neg_r \varphi))} (\to I^1)$$

$\square$

$$\frac{\begin{array}{c}[\varphi]_i\\\vdots\\\psi\end{array}}{\varphi\to\psi}\,(\to I^i) \qquad \frac{\varphi\to\psi\quad\varphi}{\psi}\,(\to E) \hfill (3.5.47)$$

$$\frac{\begin{array}{c}[\neg_l\varphi]_i\\\vdots\\\psi\end{array}}{\neg_l(\varphi\to\psi)}\,(\neg_l\to I^i) \qquad \frac{\neg_l(\varphi\to\psi)\quad\neg_l\varphi}{\psi}\,(\neg_l\to E) \hfill (3.5.48)$$

$$\frac{\begin{array}{cc}[p]_i & [p]_i\\\vdots\\ q & \neg_l q\end{array}}{\neg_l p}\,(At\neg_l I^i) \qquad \frac{p\quad\neg_l p}{\varphi}\,(At\neg_l E) \hfill (3.5.49)$$

$$\frac{\neg_l\neg_l\varphi}{\varphi}\,(dne_l) \qquad \frac{\varphi}{\neg_l\neg_l\varphi}\,(dni_l) \hfill (3.5.50)$$

Figure 3.4: The I/E-rules of $\mathcal{N}^{\neg_l}$

3.5.6 The natural-deduction system $\mathcal{N}^{\neg_l}$

The design of the ND-system $\mathcal{N}^{\neg_l}$ in Figure 3.4 is based on analogous principles to those driving $\mathcal{N}^{\neg_r}$.

Proposition 3.5.21 (closure under composition). *Derivations of $\mathcal{N}^{\neg_l}$ are closed under composition of derivations.*

Again, the proof is standard and omitted.

Corollary 3.5.8. *The mutual derivability $\neg_l(\varphi\to\psi)\dashv\vdash_{\mathcal{N}^{\neg_l}}\neg_l\varphi\to\psi$ (cf. (3.5.23)) holds.* □

3.5 Francez's Connexive logics

Proof. The $\mathcal{N}^{\neg_l}$-derivations are as follows.

$$\dfrac{\dfrac{\neg_l\varphi\to\psi \quad [\neg_l\varphi]_1}{\psi}(\to E)}{\neg_l(\varphi\to\psi)}(\neg_l\to I^1) \qquad \dfrac{\dfrac{\neg_l(\varphi\to\psi) \quad [\neg_l\varphi]_1}{\psi}(\neg_l\to E)}{\neg_l\varphi\to\psi}(\to I^1) \tag{3.5.51}$$

3.5.6.1 Derivability of the connexive axioms in $\mathcal{N}^{\neg_l}$

I show the $\mathcal{N}^{\neg_l}$-derivability of the connexive axioms the system $\mathcal{N}^{\neg_l}$, justifying its defining a Connexive logic too.

Proposition 3.5.22 (Aristotle's $\neg_l$-thesis).

$$\vdash_{\mathcal{N}^{\neg_l}} \neg_l(\neg_l\varphi\to\varphi) \tag{3.5.52}$$

Proof. The $\mathcal{N}^{\neg_l}$-derivation is as follows.

$$\dfrac{\dfrac{[\neg_l\neg_l\varphi]_1}{\varphi}[(dne_l)]}{\neg_l(\neg_l\varphi\to\varphi)}(\neg_l\to I^1)$$

Proposition 3.5.23 (Boethius' $\neg_l$-theses).

$$(B_3)\ \vdash_{\mathcal{N}^{\neg_l}} (\varphi\to\psi)\to\neg_l(\neg_l\varphi\to\psi) \tag{3.5.53}$$

$$(B_4)\ \vdash_{\mathcal{N}^{\neg_l}} (\neg_l\varphi\to\psi)\to\neg_l(\varphi\to\psi)$$

Proof. The $\mathcal{N}^{\neg_l}$-derivation of (B_3) is as follows.

$$\cfrac{\cfrac{[\varphi\to\psi]_2 \quad \cfrac{[\neg_l\neg_l\varphi]_1}{\varphi}(dne_l)}{\cfrac{\psi}{\cfrac{\neg_l(\neg_l\varphi\to\psi)}{(\varphi\to\psi)\to\neg_l(\neg_l\varphi\to\psi)}(\to I^2)}(\neg_l\to I_1^1)}(\to E)} \qquad (3.5.54)$$

The $\mathcal{N}^{\neg_l}$-derivation of (B_4) is as follows.

$$\cfrac{\cfrac{[\neg_l\varphi\to\psi]_2 \quad [\neg_l\varphi]_1}{\cfrac{\psi}{\cfrac{\neg_l(\varphi\to\psi)}{(\neg_l\varphi\to\psi)\to\neg_l(\varphi\to\psi)}(\to I^2)}(\neg_l\to I^1)}(\to E)} \qquad (3.5.55)$$

$\square$

3.5.6.2 Negation inconsistency of $\mathcal{N}^{\neg_l}$

The following observation is due to Omori [84].

Proposition 3.5.24 (Negation inconsistency of $\mathcal{N}^{\neg_l}$).

$$\begin{aligned} &\vdash_{\mathcal{N}^{\neg_l}} \varphi\to(\varphi\to\varphi) \\ &\vdash_{\mathcal{N}^{\neg_l}} \neg_l(\varphi\to(\varphi\to\varphi)) \end{aligned} \qquad (3.5.56)$$

Proof. The $\mathcal{N}^{\neg_l}$-derivations are:

$$\cfrac{\cfrac{[\varphi]_1 \quad [\varphi]_2}{\cfrac{\varphi\to\varphi}{\varphi\to(\varphi\to\varphi)}(\to I^1)}(\to I^2)} \qquad \cfrac{\cfrac{[\varphi]_1 \quad [\neg\varphi]_2}{\cfrac{\varphi\to\varphi}{\neg_l(\varphi\to(\varphi\to\varphi))}(\neg_l\to I^3)}(\to I^1)}$$

Note the vacuous discharges in both derivations. $\square$

3.5.7 Model-theory for $\mathcal{N}^{\neg r}$ and $\mathcal{N}^{\neg l}$

In this section, I present a model-theory for the two ND proof-systems $\mathcal{N}^{\neg r}$ and $\mathcal{N}^{\neg l}$ presented above. The role of the model-theory is merely a tool for establishing indirectly some properties of the ND-systems, such as *non-derivability*.

The model-theory of both systems is based on a *four-valued* truth-tables, having the the truth-values $\mathcal{V} = \{0, 1, 2, 3\}$. As an intuitive handle to the interpretation of those values, one can think of them as binary representations of *ordered pairs* of classical truth-values $\{0, 1\}$. Let the *designated* values be $D = \{0, 1, 3\}$.

A characteristic fact of the definition of the two negations is their *non-compositionality*: they are not truth-functional in that the truth-value of a negated conditional does not depend on the truth-value of the conditional itself; rather, it depends on the truth-values of both the antecedent and the consequent of the conditional, coded as an ordered pair of classical values. Thus, the value 2, coding $\langle 1, 0 \rangle$, is the falsity in this system, coding the value of a conditional with a true antecedent and a false consequent. In that, the model-theory reflects the *non-purity* of the negated-conditional rules in both systems. The other truth-values are variants of classical truth, recording *in virtue of which combination of values for the consequent and the antecedent is the conditional true*.

When considering valuations (truth-value assignments), atomic propositions differ from compound propositions (conditionals): the former are assigned *only* $\{2, 3\}$, rendering them contradictory.

Definition 3.5.3 (valuation). *A valuation* $v : \text{At} \mapsto \mathcal{V}$ *is a mapping satisfying*

$$\begin{cases} v[\![p]\!] \in \{2, 3\} & atomic \\ v[\![\varphi]\!] \in \{0, 1, 2, 3\} & compound \end{cases}$$

Thus, neither of the logics is closed under the rule of uniform substitu-

p	$\neg_r p$	$\neg_l p$
2	3	3
3	2	3

φ	$\neg_r \varphi$	$\neg_l \varphi$
3	1	2
2	0	3
1	3	0
0	2	1

Figure 3.5: The truth-tables for $\neg_r$ and $\neg_l$

tion. This reflects an intuition coming from natural language, that the generators of the object-language are *atomic propositions*, having unspecified but fixed contents, and not propositional variables! A similar phenomenon, similarly justified, takes place also in the system OL of Cooper [14].

The truth-tables for the two negations are presented in Figure 3.5. The truth-table of the conditional is presented in Figure 3.6. The line marked with ($*$) are the falsity lines, yielding false for a false consequent and the various ways the antecedent can be true (designated).

Definition 3.5.4 (equivalence). *Let $\varphi \equiv \psi$ iff for every valuation v:* $v[\![\varphi]\!] = v[\![\psi]\!]$.

□

3.5 Francez's Connexive logics

φ	ψ	$\varphi\to\psi$	
3	3	3	
3	2	2	(∗)
3	1	3	
3	0	3	
2	3	1	
2	2	0	
2	1	1	
2	0	1	
1	3	3	
1	2	2	(∗)
1	1	3	
1	0	3	
0	3	3	
0	2	2	(∗)
0	1	3	
0	0	3	

Figure 3.6: The truth-table for →

In those systems, *soundness* of a rule means that for every valuation v: if the premises of the rule are assigned a designated truth-value under v, so does the conclusion. A simple case analysis establishes the following proposition.

Proposition 3.5.25 (soundness). $\mathcal{N}^{\neg r}$ *and* $\mathcal{N}^{\neg l}$ *are sound w.r.t. the model-theory.*

Proof. The proof is by case analysis. As an example, consider the rule $(\neg_l \to I)$ (in Figure 3.4). In order to lead from a designated truth-value for $\neg_l \varphi$ to a designated truth-value of ψ, the possible truth-values of φ are $\{1, 0, 2\}$ and the possible truth-values of ψ are $\{0, 1, 3\}$. For each combination of the above, the truth-value of $\neg_l(\varphi \to \psi)$ is 0, designated too.

Similarly, the following equivalences (identical truth-values) justify (3.5.23).
$$\neg_r(\varphi \to \psi) \equiv \varphi \to \neg_r \psi \qquad \neg_l(\varphi \to \psi) \equiv \neg_l \varphi \to \psi$$
These are the hyper-connexivity axioms relativized to each negation separately. □

3.5.7.1 Non-derivability of some classical laws

Example 3.5.14 (non-derivability of contraposition). *As an example of non-derivability, both negations invalidate* contraposition *(as might be expected, cf. Section 2.5.3).*

- $\neg_r$: Consider the valuation v' under which $v'[\![\psi]\!] = 2$ and $v'[\![\varphi]\!] = 3$. Hence, $v'[\![\neg_r \psi]\!] = 3$ and $v'[\![\neg_r \varphi]\!] = 2$. Therefore, $v'[\![\neg_r \psi \to \neg_r \varphi]\!] = 2$ *(false!)*, while $v'[\![\varphi \to \psi]\!] = 3$ *(true!)*.

- $\neg_l$: Consider the valuation v'' under which $v''[\![\psi]\!]i = 1$ and $v''[\![\varphi]\!] = 0$. Hence, $v''[\![\neg_l \psi]\!] = 3$ and $v''[\![\neg_l \varphi = 2]\!]$. Therefore, $v''[\![\neg_l \psi \to \neg_l \varphi]\!] = 2$ *(false!)*, while $v''[\![\varphi \to \psi]\!] = 3$ *(true!)*.

3.5 Francez's Connexive logics

□

Contraposition does hold for atomic sentences, for which such valuations are inadmissible. The same situation obtains also in the system OL of Cooper [14].

As another example, I show non-explosion.

Example 3.5.15 (**non-explosiveness**). *Consider a valuation v^* s.t. $v^*[\![\varphi]\!] = v^*[\![\psi]\!] = 3$. Thus, both $v^*[\![\neg_l(\varphi \to \psi)]\!]$ $[= v^*[\![\neg_l\varphi \to \psi]\!]] = 3$ (truth!), and $v^*[\![\varphi \to \psi]\!] = 3$ (truth), but an arbitrary ξ can certainly have $v^*[\![\xi]\!] = 2$ (falsity!). Thus,*

$$\neg_l(\varphi \to \psi), (\varphi \to \psi) \not\models \xi$$

A similar argument applies to '$\neg_r$'.

□

Example 3.5.16 (**Modus Tollens**). *Neither of the two versions of the modus tollens rule are validated.*

$$\frac{\varphi \to \psi \quad \neg_r \psi}{\neg_r \varphi} \ (MT_r) \qquad \frac{\varphi \to \psi \quad \neg_l \psi}{\neg_l \varphi} \ (MT_l) \qquad (3.5.57)$$

- *A counter example for (MT_r): consider a valuation v with $v[\![\varphi]\!] = 0$ and $v[\![\psi]\!] = 3$. Therefore, $v[\![\varphi \to \psi]\!] = 3$ (designated), and $v[\![\neg_r \psi]\!] = 1$ (designated); however, $v[\![\neg_r \varphi]\!] = 2$ (non-designated).*

- *A counter example for (MT_l): consider a valuation v with $v[\![\varphi]\!] = 3$ and $v[\![\psi]\!] = 1$. Therefore, $v[\![\varphi \to \psi]\!] = 3$ (designated), and $v[\![\neg_l \psi]\!] = 0$ (designated); however, $v[\![\neg_l \varphi]\!] = 2$ (non-designated).*

3.5.8 Conclusion

In this section, I have introduced two negations, interacting with the conditional in a manner similar to the interaction in some known Connexive Logics. This interaction is inspired by a similar interaction present in some natural language dialogs, where negation is used to disagree about sufficiency or necessity of conditions, "correcting" the negated proposition by an alternative proposition, in which either the antecedent or the consequent of the negated implication are negated themselves.

Two ND proof-systems are proposed for the resulting logics, viewed as meaning-conferring, in accordance with the proof-theoretic semantics program. For an overview of this theory of meaning see Francez [32].

In addition, a four-valued model-theory is developed as a tool for establishing non-derivability.

An interesting alternative for capturing the same intuition might be keeping *one* negation, but splitting the conditional.

In Section 4.3 there is an extension of the approach presented here, by adding other connectives, where again negation applies by negating one argument of a binary connective. For conjunction and disjunction, this would produce an invalidation of De Morgan's rules.

Negating quantified propositions by such negations is, of course, of interest too and is left for future research.

3.6 Some more Connexive logics

In this section I mention briefly some additional Connexive logics presented in the literature, without going into detailed presentation. The reader is encouraged to study the original presentations of those logics to encounter additional issues related to connexivity.

3.6.1 Nelson

In [79], Nelson argues against the truth-functionality of the connectives, in particular the conditional, and proposes *intensional*, content based, definitions for them. His point of departure is a primitive intensional relation of *compatibility* ('lack of conflict') $\varphi \circ \psi$ between φ and ψ.

His definition of an intensional conditional is

$$\varphi \to \psi =^{df.} \neg(\varphi \circ \neg\psi)$$

i.e., the antecedent is not incompatible with the consequent.

The connexive axioms A_1, A_2, B_1 and B_2 are all formal theses of Nelson's logic NL^8. For conjunction, his definition invalidates the simplification law $\varphi \wedge \psi \to \varphi$.

A comprehensive discussion can be found in Mares and Paoli [68].

3.6.2 McCall

In [71], McCall introduces (axiomatically) a Connexive logic $CC1$, based on Angell's PA_1 truth-tables in Figure 3.2. He adds a few axioms to PA_1 and obtains an axiomatization complete for those truth-tables.

3.6.3 Pizzi and Williamson

In [93], Pizzi and Williamson investigate Connexive logics arising by adding connexive axioms to logics having a conditional known as *consequential conditional*, introduced by Pizzi in previous papers, a conditional related to modal operators. In particular, they show the connection of negating the conditional as in (2.4.17) and satisfaction of a

---
[8] A name given by Mares and Paoli [68].

related property, the *conditional excluded middle* scheme

$$(\varphi \to \psi) \vee (\varphi \to \neg \psi)$$

3.6.4 Estrada-González and Ramírez Cámara

In [26], Estrada-González and Ramírez Cámara consider a three-valued Connexive logic MRS^P, originally introduced by Estrada-González [23], not devised to satisfy the connexive axioms; rather, the motivation is modeling the evaluation of conditionals with a false antecedent and satisfying more relations in the square of opposition.

3.6.5 Wansing and Unterhuber

In [124], Wansing and Unterhuber present a Connexive logic cCL, obtained as an extension of four-valued *First-Degree Entailment (FDE)* logic, delineated in Section 7.4.1.2. This logic is further extended, in different ways, to obtain a family of *Conditional connexive logics*.
An interesting feature in the model-theoretic definition of conditionals is the use of Chellas frames, that have an accessibility relation indexed by the truth-set of the antecedent of the conditional.

3.6.6 Omori

In [83], Omori defines a Connexive logic[9] *dLP* (dialetheic *LP*), expanding Priest's logic of paradox *LP* and employing a weaker falsification

---

[9]Omori's motivation, though, was not connexivity; rather, to have a paraconsistent logic expressive enough to contain a dialetheia already at the propositional level.

3.6 Some more Connexive logics

condition for the conditional (according to Wansing's method for obtaining connexivity). Recall that in LP, the conditional $\varphi \to \psi$ is defined as $\neg\varphi \lor \psi$.

For a model-theoretic definition, let ρ be the relational valuation relating formulas to $\{1,0\}$. Then,

$$
\begin{array}{ll}
\text{Logic} & (\varphi \to \psi)\rho o \text{ iff} \\
\hline
LP & \varphi\rho 1 \text{ and } \psi\rho 0 \\
dLP & \varphi \not\rho 1 \text{ or } \psi\rho 0
\end{array}
\tag{3.6.58}
$$

In other words, the falsification condition of the conditional is:

If $\varphi\rho 1$ then $\psi\rho 0$

In the axiomatic definition, there is the negating the conditional axiom (2.4.17).

All the four connexive axioms A_1, A_2, B_1 and B_2 are validated by dLP. dLP is (negation)-inconsistent, as

$$\vdash_{dLP} (\varphi \land \neg\varphi \to \psi) \land \neg(\varphi \land \neg\varphi \to \psi)$$

3.6.7 Rahman and Rückert

In [99], Rahman and Rückert present a *Dialogical Connexive logic*.

The dialogical approach for defining logics originates from work by Lorentzen [62], and later work by Lorentzen and Lorenz [63]. It is based on presenting a logic as a *game* between two participants: a *proponent*, attempting to verify a given formula, and an *opponent*, attempting to falsify the given formula.

Each of the participants has allowable *moves*, depending on the operators of the logic being defined. The opponent can *attack* a formula, while the proponent can *defend* a formula. In addition, there are *structural rules*, controlling the order of moves.

The validity of the formula φ is defined in terms of existence of a *winning strategy* for the proponent defending φ.

The connexivity of the conditional is obtained by extending the object-language with two *reflective* operators:

- **V**: *defensibility*.

- **F**: *attack-ability*.

The moves for the connexive conditional are formulated in terms of those additional operators. For example, a proponent defending $\varphi \rightarrow \psi$ might be asked by the opponent to defend φ itself (in a sub-dialogue), excluding vacuous implication.

An important property of the logic defined this way is the invalidation of *uniform substitution*. The reader is referred to [99] for the finer details.

3.6.7.1 Jarmużek and Malinowski

In [49], Jarmużek and Malinowski introduce a family of *Boolean Connexive logics*, defined via *relatedness semantics* as defined by Jarmużek and Kaczkowski [50].

According to this semantics, an *arbitrary* binary relation R is imposed on formulas of the object-language, a relation used, usually as a filter, in defining a logic model-theoretically. By tuning R, one obtains the desired logic. Let $\not{R}$ indicate unrelatedness via R.

To obtain Connexive logics over the object-language of Classical logic, the following conditions are imposed on R:

3.6 Some more Connexive logics

- $\varphi \mathrel{R} \neg\varphi, \neg\varphi \mathrel{R} \varphi$.
- $\varphi \mathrel{R} \psi$ or $\varphi \mathrel{R} \neg\psi$.
- $\varphi \to \psi \mathrel{R} \neg(\varphi \to \neg\psi)$.
- $\varphi \to \neg\psi \mathrel{R} \neg(\varphi \to \psi)$.

Validity of $\varphi \to \psi$ is then defined by strengthening the classical validity with $\varphi R \psi$.

By varying the combinations of the above requirements on R, the different logics in the family are obtained.

In [66] Malinowski and Palczewski apply a Boolean Connexive logic for resolving and old paradox by Lewis Carroll.

Chapter 4

The scope of connexivity

4.1 Introduction

The discussion of Connexive logics so far was under the assumption that connexivity relates a conditional and a negation, where the antecedent and the consequent of the conditional are *any* formulas of the object-language.

In this chapter, I consider two deviations from this assumption.

restricting the scope of connexivity: Keeping connexivity as relating *only* a conditional and a negation, but restricting the form of the antecedent and the consequent of a valid connexive conditional.

extending the scope of connexivity: Relieving the common view that connexivity relates *only* a conditional and a negation, allowing also other connectives to interact with a negation in a way to be considered connexive too. In a sense, this restores and enhances the Stoic view of connexivity as resulting from a (vague notion of) "connection", making the latter applicable to arguments of any (binary) connective.

Clearly, such a view of connexivity abandons the orthodox view, prevailing since the modern revival of Connexive logic by Angell and McCall, that connexivity relates only a conditional and a negation.

4.2 Restricting the scope of connexivity: Humble connexivity

In [56], Kapsner raises[1] the claim that requiring the connexive axioms to hold for *any* antecedent and consequent of a connexive conditional is "too stringent a requirement". A similar opinion is expressed by Iacona [48] and Crupi and Iacona [15].

What could constitute an instance of, say, Aristotle's axioms, the plausibility of which seems less compelling?

Kapsner's answer ([56, p. 6]) is: instances with an *impossible*, or at least *contradictory*, antecedents[2]. I will consider here only the latter, deferring the former to Section 7.2 . The reason for this deferral is that considering the former requires the assumption that the object-language includes modal operators. I believe that connexivity should be discussed also *without* such an assumption.

For example, an inconsistent (self-contradictory) antecedent in an in-

---

[1] Actually, this claimed is *revived* by Kapsner, seemingly lurking around from antiquity.

[2] Crupi and Iacona [15] raise also the dual case of a necessary, or at least tautological, consequent. Since the arguments are similar, I concentrate on the former, pertaining to the antecedent.

4.2 Restricting the scope of connexivity: Humble connexivity

stance of A_1 might look like

$$\vdash \neg(\chi \wedge \neg \chi \rightarrow \neg(\chi \wedge \neg \chi)) \tag{4.2.1}$$

For example,

> It is not the case that if it is raining and it is not raining,
>
> then it is not the case that it is raining and it is not raining

Similarly, for A_2:

$$\vdash \neg(\neg(\varphi \vee \neg \varphi) \rightarrow (\varphi \vee \neg \varphi)) \tag{4.2.2}$$

Thus, Kapsner proposes to reformulate the A_i axioms as

$$\text{For any satisfiable } \varphi: \quad A_1^h: \neg(\varphi \rightarrow \neg \varphi) \quad A_2^h: \neg(\neg \varphi \rightarrow \varphi) \tag{4.2.3}$$

The same restriction of non-contradiction is suggested to be imposed on both φ and $\varphi \rightarrow \psi$ to obtain the B_i^h axioms, for the sake of their plausibility.

A similar restriction is implicit in Vidal's discussion in [117]. However, Vidal goes a step further, suggesting that contradictory antecedents should be banned from conditionals *altogether*, not just connexive ones. I find Vidal's view more tenable than restricting the connexive axioms (only). This view is an internalization into the object-language of the null account of logical consequence from a contradiction as discussed on p. 11.

One way Kapsner indicates[3] to interpret the implausibility of the theoremhood of (4.2.1) is by appealing to the *explosion* principle, whereby everything is implied by a contradiction, thereby rendering the negation of this instance of A_1, namely $\chi \wedge \neg \chi \rightarrow \neg(\chi \wedge \neg \chi)$ – true.

In the context of connexivity, within the general revolt agains Classical logic, this is definitely an unsatisfactory way.

---

[3] A way attributed to David Mackinson.

Another way, also appealing to classical principle (though less objectionable than explosion) is the following derivation[4] (my rearrangement) by McCall ([71, p. 463]).

$$\frac{\dfrac{\dfrac{\dfrac{[\chi \wedge \neg \chi]_1}{\chi}\,(\wedge E)}{\chi \vee \neg \chi}\,(\vee I)}{\neg(\chi \wedge \neg \chi)}\,(DM)}{(\chi \wedge \neg \chi) \to \neg(\chi \wedge \neg \chi)}\,(\to I^1)$$

(where(DM) is the classically derived De-Morgan rule.)

There are good grounds to reject at least ($\wedge E$) and ($\vee I$), a matter discussed in Section 2.5.

4.3 Extending the scope of connexivity: Poly-connexivity

4.3.1 Introduction

This section is based on Francez [36], where I argue in favor of *extending* the scope of connexivity, presenting a *poly-connexive* logic $PCON$ that includes also connexive conjunction and disjunction (referred to as connexive 'coordination connectives') in its object-language. The idea of extending connexivity to other connectives besides the conditional is mentioned in passing also in Rahman and Rückert [99], in the context of a Dialogical Connexive logic. They view conjunction as defined dialogically already connexive, and extend connexivity to disjunction in a manner unrelated to what follows.

---

[4]McCall presents the derivation in his axiomatic system. I find the natural-deduction presentation clearer.

4.3 Extending the scope of connexivity: Poly-connexivity

The motivation for the extension of the scope of connexivity is, that like the conditional, the coordination connectives have uses in NL exhibiting *a connection between the their two arguments*. This is explained in Section 7.6.5.

Digression: terminology
One may object to the classification of $PCON$ as a 'poly-connexive logic', suggesting instead classifying the logic as 'poly-contra-classical'. The argument is that connexivity is a property of a conditional (interacting with negation) and not of other connectives. The proposal of $PCON$ can be understood as extending Wansing's method of changing falsification conditions from the conditional to conjunction and disjunction. However, as stressed above, the extension from the conditional to the other connectives is the requirement of *a connection in content* between the two arguments of binary connectives. Changing the falsification condition is just a method to induce such a connection. The whole point of introducing $PCON$ is challenging the orthodoxy on connexivity by extending its scope to connectives other than the conditional when interacting with negation.
end of digression

The following are the characteristic interactions of the binary coordination connectives with the negation.

1.
$$(neg_i): \vdash_{PCON} \neg(\varphi \to \psi) \leftrightarrow [(\varphi \to \neg \psi) \vee (\neg \varphi \to \psi)] \quad (4.3.4)$$

2.
$$(neg_c): \vdash_{PCON} \neg(\varphi \wedge \psi) \leftrightarrow [(\varphi \wedge \neg \psi) \vee (\neg \varphi \wedge \psi)] \quad (4.3.5)$$

3.
$$(neg_d): \vdash_{PCON} \neg(\varphi \vee \psi) \leftrightarrow \neg \varphi \vee \neg \psi \quad (4.3.6)$$

Clearly, none of the above is a thesis of Classical logic.

4.3.2 Axiomatic definition of $PCON$

The object-language $\mathcal{L}_{PCON}$ consists of the usual closure of a countable set **At** of atomic propositions w.r.t. the traditional operators $\{\neg, \rightarrow, \wedge, \vee\}$. The notation $\varphi \leftrightarrow \psi$ is a shorthand for two axioms: $\varphi \rightarrow \psi$ and $\psi \rightarrow \varphi$.

Definition 4.3.5 (Axiomatic definition of $PCON$). *The Hilbert-style axiomatic definition of $PCON$ is given by the following axioms, divided into two groups, positive axioms (as for positive propositional intuitionistic logic), and negative axioms:*

positive axioms:

(**Ax1**) $\varphi \rightarrow (\psi \rightarrow \varphi)$

(**Ax2**) $(\varphi \rightarrow (\psi \rightarrow \chi)) \rightarrow ((\varphi \rightarrow \psi) \rightarrow (\varphi \rightarrow \chi))$

(**Ax3**), (**Ax4**) $\varphi \wedge \psi \rightarrow \varphi$, $\varphi \wedge \psi \rightarrow \psi$

(**Ax5**) $\varphi \rightarrow (\psi \rightarrow \varphi \wedge \psi)$

(**Ax6**), (**Ax7**) $\varphi \rightarrow \varphi \vee \psi$, $\psi \rightarrow \varphi \vee \psi$

(**Ax8**) $((\varphi \rightarrow \chi) \rightarrow ((\psi \rightarrow \chi) \rightarrow (\varphi \vee \psi) \rightarrow \chi))$

negative axioms:

(**Ax9**) $\neg\neg\varphi \leftrightarrow \varphi$

(**Ax10**) $\neg(\varphi \wedge \psi) \leftrightarrow [(\varphi \wedge \neg\psi) \vee (\neg\varphi \wedge \psi)]$

(**Ax11**) $\neg(\varphi \vee \psi) \leftrightarrow (\neg\varphi \vee \neg\psi)$

(**Ax12**) $\neg(\varphi \rightarrow \psi) \leftrightarrow [(\varphi \rightarrow \neg\psi) \vee (\neg\varphi \rightarrow \psi)]$

with the single inference rule Modus Ponens (MP)

$$\frac{\varphi \quad \varphi \rightarrow \psi}{\psi} \; (MP)$$

4.3 Extending the scope of connexivity: Poly-connexivity 109

Notably, *closure under uniform substitution* is *not* a rule in $PCON$.

For a finite Γ, *derivability* in $PCON$ of φ from Γ, denote by $\Gamma \vdash_{PCON} \varphi$, is defined as usual. The proof of the following proposition is standard and omitted.

Proposition 4.3.26 (deduction theorem).

$$\Gamma, \varphi \vdash_{PCON} \psi \text{ iff } \Gamma \vdash_{PCON} \varphi \to \psi \qquad (4.3.7)$$

4.3.3 Models for $PCON$

The following relational models for $PCON$ are obtained by suitable modifications of those in Nelson's logic **N4** (see Kamide and Wansing [52]). The formulation follows Omori [84].

Definition 4.3.6 (models for $PCON$). *A model for $PCON$ is a triple $\langle W, \leq, V \rangle$, where:*

- *W is a non-empty set (of points or states)*
- *'$\leq$' is a partial-order on W*
- *$V : W \times \mathbf{At} \Rightarrow \{\varnothing, \{0\}, \{1\}, \{0,1\}\}$ is an assignment of truth values to pairs of states and atomic propositions with the condition that $i \in V(w1, p)$ and $w1 \leq w2$ only if $i \in V(w2, p)$ for all $p \in \mathbf{At}$, all $w1, w2 \in W$ and $i \in \{0, 1\}$.*

Valuations V are then extended to interpretations I to state-formula pairs by the following conditions:

- *$I(w, p) = V(w, p)$*
- *$1 \in I(w, \neg\varphi)$ iff $0 \in I(w, \varphi)$*

- $0 \in I(w, \neg\varphi)$ iff $1 \in I(w, \varphi)$

- $1 \in I(w, \varphi\wedge\psi)$ iff $1 \in I(w, \varphi)$ and $1 \in I(w, \psi)$

- $0 \in I(w, \varphi\wedge\psi)$ iff *one of the following two conditions holds.*
    - either $1 \in I(w, \varphi)$ and $0 \in I(w, \psi)$
    - or $0 \in I(w, \varphi)$ and $1 \in I(w, \psi)$

- $1 \in I(w, \varphi\vee\psi)$ iff $1 \in I(w, \varphi)$ or $1 \in I(w, \psi)$

- $0 \in I(w, \varphi\vee\psi)$ iff $0 \in I(w, \varphi)$ or $0 \in I(w, \psi)$.

- $1 \in I(w, \varphi\to\psi)$ iff for all $x \in W$: if $w \leq x$ and $1 \in I(x, \varphi)$ then $1 \in I(x, \psi)$

- $0 \in I(\varphi\to\psi)$ iff *one of the following two conditions holds.*
    - either *for all* $x \in W$ s.t. $w \leq x$: if $1 \in I(x, \varphi)$ then $0 \in I(x, \psi)$,
    - or *for all* $x \in W$ s.t. $w \leq x$: if $0 \in I(x, \varphi)$ then $1 \in I(x, \psi)$

Definition 4.3.7 (logical consequence). $\Sigma\vDash_{PCON} \varphi$ *iff for all models* $\langle W, \leq, I\rangle$, *and for all* $w \in W$: $1 \in I(w, \varphi)$ *if* $1 \in I(w, \psi)$ *for all* $\psi \in \Sigma$.

4.3.3.1 Classical validities invalidated by $PCON$

Proposition 4.3.27 (invalidating De Morgan's laws).

$$\nvDash_{PCON} \neg(\varphi\wedge\psi) \leftrightarrow \neg\varphi\vee\neg\psi \tag{4.3.8}$$

Proof. A counter-model to the instance of De Morgan's law $\neg(p\wedge q) \leftrightarrow \neg p\vee\neg q$ has one w, with $0 \in I(w, p)$ and $0 \in I(w, q)$. Therefore, $1 \in I(w, \neg p)$ and $1 \in I(w, \neg q)$, implying $1 \in I(w, \neg p\vee\neg q)$. On the other hand, $1 \notin I(w, \neg(p\wedge q))$ since $0 \in I(w, \neg(p\wedge q))$. □

Proposition 4.3.28 (invalidating modus tollens).

$$\varphi \to \psi, \neg\psi \not\vdash_{PCON} \neg\varphi \qquad (4.3.9)$$

Proof. A counter-model again contains a single w with:

- $1 \in I(w,p)$, $0 \notin I(w,p)$
- $\{0,1\} \subseteq I(w,q)$

Hence, $1 \in I(w, p \to q)$ and $1 \in I(w, \neg q)$, but $1 \notin I(w, \neg p)$. □

4.3.4 The rules of $\mathcal{N}_{PCON}$

The system draws on ideas from Francez [33], where negation was split. $\mathcal{N}_{PCON}$ is presented in Figures 4.1 and 4.2. The first figure presents the standard positive rules. The novel part, the negative rules (corresponding to falsification) are presented in Figure 4.2.

Remarks on the rules:

conjunction: The rules express the following equivalence, induced by the falsification condition of $\neg(\varphi \wedge \psi)$:

$$\neg(\varphi \wedge \psi) \equiv (\varphi \wedge \neg\psi) \vee (\neg\varphi \wedge \psi)$$

disjunction: Here the equivalence induced by the falsification condition of $\neg(\varphi \vee \psi)$, reflected by the rules, is:

$$\neg(\varphi \vee \psi) \equiv \neg\varphi \vee \neg\psi$$

$$\frac{\begin{array}{c}[\varphi]_i\\ \vdots\\ \psi\end{array}}{\varphi\to\psi}\,(\to I^i) \qquad \frac{\varphi\to\psi\quad\varphi}{\psi}\,(\to E) \hfill (4.3.10)$$

$$\frac{\varphi\quad\psi}{\varphi\wedge\psi}\,(\wedge I) \qquad \frac{\varphi\wedge\psi}{\varphi}\,(\wedge E_1) \qquad \frac{\varphi\wedge\psi}{\psi}\,(\wedge E_2) \hfill (4.3.11)$$

$$\frac{\varphi}{\varphi\vee\psi}\,(\vee I_1) \qquad \frac{\psi}{\varphi\vee\psi}\,(\vee I_2) \qquad \frac{\varphi\vee\psi\quad\begin{array}{c}[\varphi]_i\\ \vdots\\ \chi\end{array}\quad\begin{array}{c}[\psi]_j\\ \vdots\\ \chi\end{array}}{\chi}\,(\vee E^{i,j}) \hfill (4.3.12)$$

Figure 4.1: $\mathcal{N}_{PCON}$: the positive fragment

implication: The equivalence induced by the falsification condition of $\neg(\varphi\to\psi)$ is
$$\neg(\varphi\to\psi) \equiv (\varphi\to\neg\psi)\vee(\neg\varphi\to\psi)$$
There is a certain complication in the structure of the (E)-rule, that has premises of the form *discharged sub-derivations* (that can be thought of as *discharged rules*). This complication arises due to the fact that there are *two* (I)-rules, *both of which* discharge assumptions. See Schroeder-Heister [108, 110] and Olkhovikov and Schroeder-Heister [82] for the need of such rules and the circumstances leading to them.

Tree-shaped *derivations* are defined recursively, almost as usual. The only difference is that an assumed rule may be applied, and discharged immediately after application, in addition to the applications of primitive rules. Examples of derivations are presented below, in the proof of deductive equivalence of the axiomatic presentation of $PCON$ and its ND-presentation, $\mathcal{N}_{PCON}$.

4.3 Extending the scope of connexivity: Poly-connexivity

$$\frac{\varphi \quad \neg\psi}{\neg(\varphi\wedge\psi)}\,(\neg\wedge I_1) \qquad \frac{\neg\varphi \quad \psi}{\neg(\varphi\wedge\psi)}\,(\neg\wedge I_2)$$

$$\frac{\neg(\varphi\wedge\psi) \quad \begin{matrix}[\varphi,\neg\psi]_i\\ \vdots\\ \chi\end{matrix} \quad \begin{matrix}[\neg\varphi,\psi]_j\\ \vdots\\ \chi\end{matrix}}{\chi}\,(\neg\wedge E^{i,j})$$
(4.3.13)

$$\frac{\neg\varphi}{\neg(\varphi\vee\psi)}\,(\neg\vee I_1) \qquad \frac{\neg\psi}{\neg(\varphi\vee\psi)}\,(\neg\vee I_2)$$

$$\frac{\neg(\varphi\vee\psi) \quad \begin{matrix}[\neg\varphi]_i\\ \vdots\\ \chi\end{matrix} \quad \begin{matrix}[\neg\psi]_j\\ \vdots\\ \chi\end{matrix}}{\chi}\,(\neg\vee E^{i,j})$$
(4.3.14)

$$\frac{\begin{matrix}[\varphi]_i\\ \vdots\\ \neg\psi\end{matrix}}{\neg(\varphi\to\psi)}\,(\neg\to I_1^i) \qquad \frac{\begin{matrix}[\neg\varphi]_i\\ \vdots\\ \psi\end{matrix}}{\neg(\varphi\to\psi)}\,(\neg\to I_2^i)$$

$$\frac{\neg(\varphi\to\psi) \quad \begin{matrix}\begin{bmatrix}\varphi\\ \vdots\\ \neg\psi\end{bmatrix}_i\end{matrix} \quad \begin{matrix}\begin{bmatrix}\neg\varphi\\ \vdots\\ \psi\end{bmatrix}_j\end{matrix}}{\chi}\,(\neg\to E^{i,j})$$
(4.3.15)

$$\frac{\neg\neg\varphi}{\varphi}\,(dne) \qquad \frac{\varphi}{\neg\neg\varphi}\,(dni)$$
(4.3.16)

Figure 4.2: $\mathcal{N}_{PCON}$: the negative fragment

4.3.5 Connexivity of $PCON$

The first two examples below establish the $PCON$-derivability of the A_i and B_i axioms.

Example 4.3.17 (A_i, $i = 1, 2$).

$$\cfrac{\cfrac{[\varphi]_1}{\neg\neg\varphi}\ (dni)}{\neg(\varphi\to\neg\varphi)}\ (\neg\to I_1^1) \qquad \cfrac{\cfrac{[\neg\neg\varphi]_1}{\varphi}\ (dne)}{\neg(\neg\varphi\to\varphi)}\ (\neg\to I_2^1) \qquad (4.3.17)$$

Example 4.3.18 (B_i $i = 1, 2$).

$$\cfrac{\cfrac{\cfrac{[\varphi\to\psi]_1\ \ [\varphi]_2}{\psi}\ (\to E)}{\cfrac{\neg\neg\psi}{\neg(\varphi\to\neg\psi)}\ (\neg\to I_1^2)}\ (dni)}{(\varphi\to\psi)\to\neg(\varphi\to\neg\psi)}\ (\to I^1) \qquad \cfrac{\cfrac{\cfrac{[\varphi\to\neg\psi]_1\ \ [\varphi]_2}{\neg\psi}\ (\to E)}{\neg(\varphi\to\psi)}\ (\neg\to I_1^2)}{(\varphi\to\neg\psi)\to\neg(\varphi\to\psi)}\ (\to I^1) \qquad (4.3.18)$$

The next examples deal with the axioms for connexive conjunction and disjunction.

Example 4.3.19.

$$\cfrac{\cfrac{\cfrac{[\varphi]_j\ \ [\neg\psi]_j}{\varphi\wedge\neg\psi}\ (\wedge I)}{[\neg(\varphi\vee\psi)]_i\ \ \cfrac{}{(\varphi\wedge\neg\psi)\vee(\neg\varphi\wedge\psi)}}\ (\vee I_1) \quad \cfrac{\cfrac{[\neg\varphi]_k\ \ [\psi]_k}{\neg\varphi\wedge\psi}\ (\wedge I)}{(\varphi\wedge\neg\psi)\vee(\neg\varphi\wedge\psi)}\ (\vee I_2)}{\cfrac{(\varphi\wedge\neg\psi)\vee(\neg\varphi\wedge\psi)}{\neg(\varphi\wedge\psi)\to((\varphi\wedge\neg\psi)\vee(\neg\varphi\wedge\psi))}\ (\to I^i)}\ (\neg\vee I^{j,k})$$

$$(4.3.19)$$

$$\cfrac{[(\varphi\wedge\neg\psi)\vee(\neg\varphi\wedge\psi)]_i\ \ \cfrac{\cfrac{[\varphi\wedge\neg\psi]_j}{\varphi}\ (\wedge E_1)\ \ \cfrac{[\varphi\wedge\neg\psi]_j}{\neg\psi}\ (\wedge E_2)}{\varphi\wedge\neg\psi}\ (\wedge I)\ \ \cfrac{\cfrac{[\neg\varphi\wedge\psi]_j}{\neg\varphi}\ (\wedge E_1)\ \ \cfrac{[\neg\varphi\wedge\psi]_j}{\psi}\ (\wedge E_2)}{\neg\varphi\wedge\psi}\ (\wedge I)}{\cfrac{\neg(\varphi\wedge\psi)}{((\varphi\wedge\neg\psi)\vee(\neg\varphi\wedge\psi))\to\neg(\varphi\wedge\psi)}\ (\to I^i)}\ (\vee E^{j,k})$$

$$(4.3.20)$$

4.3 Extending the scope of connexivity: Poly-connexivity

Example 4.3.20.

$$\cfrac{[\neg(\varphi\vee\psi)]_i \quad \cfrac{\cfrac{[\neg\varphi]_j}{\neg\varphi\vee\neg\psi}(\vee I_1) \quad \cfrac{[\neg\psi]_j}{\neg\varphi\vee\neg\psi}(\vee I_1)}{\neg\varphi\vee\neg\psi}(\neg\vee E^{j,k})}{\cfrac{\neg\varphi\vee\neg\psi}{\neg(\varphi\vee\psi)\rightarrow(\neg\varphi\vee\neg\psi)}(\rightarrow I^i)} \quad (4.3.21)$$

$$\cfrac{[\neg\varphi\vee\neg\psi] \quad \cfrac{\cfrac{[\neg\varphi]_i}{\neg(\varphi\vee\psi)}(\neg\vee I_1) \quad \cfrac{[\neg\psi]_i}{\neg(\varphi\vee\psi)}(\neg\vee I_2)}{\neg(\varphi\vee\psi)}(\neg\vee E^{j,k})}{\cfrac{\neg(\varphi\vee\psi)}{(\neg\varphi\vee\neg\psi)\rightarrow\neg(\varphi\vee\psi)}(\rightarrow I^i)} \quad (4.3.22)$$

Finally, the next example establishes the new axiom for the connexive conditional.

Example 4.3.21. *Let* $\rho = \begin{bmatrix} \varphi \\ \vdots \\ \neg\psi \end{bmatrix}$ *and* $\rho' = \begin{bmatrix} \neg\varphi \\ \vdots \\ \psi \end{bmatrix}$

$$\cfrac{\neg(\varphi\rightarrow\psi) \quad \cfrac{\cfrac{\cfrac{[\varphi]_i}{\neg\psi}(\langle\rho\rangle_n)}{\varphi\rightarrow\neg\psi}(\rightarrow I^i)}{(\varphi\rightarrow\neg\psi)\vee(\neg\varphi\rightarrow\psi)}(\vee I_1) \quad \cfrac{\cfrac{\cfrac{[\neg\varphi]_j}{\psi}(\langle\rho'\rangle_m)}{\varphi\rightarrow\neg\psi}(\rightarrow I^j)}{(\varphi\rightarrow\neg\psi)\vee(\neg\varphi\rightarrow\psi)}(\vee I_2)}{(\varphi\rightarrow\neg\psi)\vee(\neg\varphi\rightarrow\psi)}(\neg\rightarrow E^{n,m}) \quad (4.3.23)$$

Note the discharge of the assumed rules ρ *and* ρ' *by the application of* $(\neg IE)$.

4.3.6 Deductive equivalence of $PCON$ and $\mathcal{N}_{PCON}$

Theorem 4.3.1.
$$\Gamma \vdash_{PCON} \varphi \text{ iff } \Gamma \vdash_{\mathcal{N}_{PCON}} \varphi \quad (4.3.24)$$

Proof: The part of the proof concerning the positive fragment is standard and omitted. I show the proof for the negative fragment.

1. Assume $\Gamma \vdash_{\mathcal{N}_{PCON}} \varphi$. To show that $\Gamma \vdash_{PCON} \varphi$, it suffices to show the derivability in $\mathcal{N}_{PCON}$ of the negative axioms.

   (Ax9): The derivations are:

   $$\dfrac{\dfrac{[\varphi]_1}{\neg\neg\varphi}\,(dni)}{\varphi\to\neg\neg\varphi}\,(\to I_1^1) \qquad \dfrac{\dfrac{[\neg\neg\varphi]_1}{\varphi}\,(dne)}{\neg\neg\varphi\to\varphi}\,(\to I_2^1)$$

   (Ax10): The derivations are:

   $$\dfrac{(\varphi\wedge\neg\psi)\vee(\neg\varphi\wedge\psi) \qquad \dfrac{\dfrac{\dfrac{[\varphi\wedge\neg\psi]_1}{\varphi}\,(\wedge E_1) \quad \dfrac{[\varphi\wedge\neg\psi]_1}{\neg\psi}\,(\wedge E)}{\neg(\varphi\wedge\psi)}\,(\neg\vee I_1) \quad \dfrac{\dfrac{[\neg\varphi\wedge\psi]_1}{\neg\varphi}\,(\wedge E_1) \quad \dfrac{[\neg\varphi\wedge\psi]_2}{\psi}\,(\wedge E_2)}{\neg(\varphi\wedge\psi)}\,(\neg\vee I_2)}{\neg(\varphi\wedge\psi)}\,(\vee E^{1,2})}{[(\varphi\wedge\neg\psi)\vee(\neg\varphi\wedge\psi)]\to\neg(\varphi\wedge\psi)}\,(\to I_1^1)$$

   and

   $$\dfrac{[\neg(\varphi\wedge\psi)]_1 \quad \dfrac{\dfrac{[\varphi,\neg\psi]_2}{\varphi\wedge\neg\psi}\,(\wedge I)}{(\varphi\wedge\neg\psi)\vee(\neg\varphi\wedge\psi)}\,(\vee I_1) \quad \dfrac{\dfrac{[\neg\varphi,\psi]_3}{\neg\varphi\wedge\psi}\,(\wedge I)}{(\varphi\wedge\neg\psi)\vee(\neg\varphi\wedge\psi)}\,(\vee I_2)}{\dfrac{(\varphi\wedge\neg\psi)\vee(\neg\varphi\wedge\psi)}{\neg(\varphi\wedge\psi)\to[(\varphi\wedge\neg\psi)\vee(\neg\varphi\wedge\psi)]}\,(\to I^1)}\,(\neg\wedge E)$$

   (Ax11): The derivations are:

   $$\dfrac{[\neg\varphi\vee\neg\psi]_1 \quad \dfrac{\dfrac{[\neg\varphi]_2}{\neg(\varphi\vee\psi)}\,(\neg\vee I_1) \quad \dfrac{[\neg\psi]_3}{\neg(\varphi\vee\psi)}\,(\neg\vee I_2)}{\neg(\varphi\vee\psi)}\,(\vee E^{2,3})}{(\neg\varphi\vee\neg\psi)\to\neg(\varphi\vee\psi)}\,(\to I^1)$$

   and

   $$\dfrac{[\neg(\varphi\vee\psi)]_1 \quad \dfrac{\dfrac{[\neg\varphi]_2}{\neg\varphi\vee\neg\psi}\,(\vee I_1) \quad \dfrac{[\neg\psi]_3}{\neg\varphi\vee\neg\psi}\,(\vee I_2)}{(\neg\varphi\vee\neg\psi)}\,(\neg\vee E^{2,3})}{\neg(\varphi\vee\psi)\to(\neg\varphi\vee\neg\psi)}\,(\to I^1)$$

4.3 Extending the scope of connexivity: Poly-connexivity 117

(Ax12): The derivations are as follows.

$$\dfrac{[(\varphi \to \neg \psi) \vee (\neg \varphi \to \psi)]_1 \quad \dfrac{\dfrac{[\varphi \to \neg \psi]_2 \quad [\varphi]_3}{\neg \psi}(\to E) \quad \dfrac{[\neg \varphi \to \psi]_4 \quad [\neg \varphi]_5}{\psi}(\to E)}{\dfrac{\neg(\varphi \to \psi)}{\neg(\varphi \to \psi)}(\vee E^{2,4})}}{\dfrac{\neg(\varphi \to \psi)}{[(\varphi \to \neg \psi) \vee (\neg \varphi \to \psi)] \to \neg(\varphi \to \psi)}(\to I^1)}$$

For the second derivation, note the applications of assumed rules, denoted as $*_1$ and $*_2$.

$$\dfrac{[\neg(\varphi \to \psi)]_1 \quad \dfrac{\dfrac{\dfrac{[\varphi]_2}{\neg \psi}(*_1)}{\varphi \to \neg \psi}(\to I^2)}{(\varphi \to \neg \psi) \vee (\neg \varphi \to \psi)}(\vee I_1) \quad \dfrac{\dfrac{\dfrac{[\neg \varphi]_3}{\psi}(*_2)}{\neg \varphi \to \psi}(\to I^3)}{(\varphi \to \neg \psi) \vee (\neg \varphi \to \psi)}(\vee I_2)}{\dfrac{[(\varphi \to \neg \psi) \vee (\neg \varphi \to \psi)]}{\neg(\varphi \to \psi) \to [(\varphi \to \neg \psi) \vee (\neg \varphi \to \psi)]}(\to I^1)}(\to E^{*1,*2})$$

2. Assume $\Gamma \vdash_{\mathcal{N}_{PCON}} \varphi$. The proof that $\Gamma \vdash_{PCON} \varphi$ proceeds by induction on the $\mathcal{N}_{PCON}$-derivation, analyzing the last $\mathcal{N}_{PCON}$-rule applied. Again, only the cases of negative rules are shown.

$\neg \wedge I_1$: The premises for this last application $(\neg \wedge I)$ are
(1) $\Gamma \vdash_{PCON} \varphi$ and (2) $\Gamma \vdash_{PCON} \neg \psi$.
By the induction hypothesis, (3) $\Gamma \vdash_{PCON} \varphi$ and
(4) $\Gamma \vdash_{PCON} \psi$. By $(Ax5)$ and twice (MP) we get
(6) $\Gamma \vdash_{PCON} \varphi \wedge \neg \psi$. By applying $(\vee I_1)$ to (6), we get
(7) $\Gamma \vdash_{PCON} (\varphi \wedge \neg \psi) \vee (\neg \varphi \wedge \psi)$. By applying (MP) to (7) and $(Ax10)$ we get the required $\Gamma \vdash_{PCON} \neg(\varphi \wedge \psi)$.

$\neg \wedge I_2$: Similar.

$\neg \wedge E$: The premises of the rule are $\neg(\varphi \wedge \psi)$, and two sub-derivations of χ, one from $\varphi, \neg \psi$ and the other from $\neg \varphi, \psi$.
By the induction hypothesis,
(1) $\Gamma \vdash_{PCON} \neg(\varphi \wedge \psi)$, (2) $\Gamma, \varphi, \neg \psi \vdash_{PCON} \chi$ and
(3) $\Gamma, \neg \varphi, \psi \vdash_{PCON} \chi$. By applying (MP) to (1) and $(Ax10)$, we get (4) $\Gamma \vdash_{PCON} (\varphi \wedge \neg \psi) \vee (\neg \varphi \wedge \psi)$. From (2) and (3) we get, using the deduction theorem,
(5) $\Gamma \vdash_{PCON} \varphi \wedge \neg \psi \to \chi$ and (6) $\Gamma \vdash_{PCON} \neg \varphi \wedge \psi \to \chi$. Using $(Ax8)$, we get

$\Gamma \vdash_{PCON} ((\varphi \wedge \neg \psi) \vee (\neg \varphi \wedge \psi)) \to \chi$. By applying (MP) to (4) and (8), we get the required $\Gamma \vdash_{PCON} \chi$.

- $\neg \to I_1$: The premise of this rule is a sub-derivation of $\neg \psi$ from φ. By the induction hypothesis, (1) $\Gamma, \varphi \vdash_{PCON} \neg \psi$. By the deduction theorem, (2) $\Gamma \vdash_{PCON} \varphi \to \neg \psi$. By $(Ax6)$, we get (3) $\Gamma \vdash_{PCON} (\varphi \to \neg \psi) \vee (\neg \varphi \to \psi)$. By applying (MP) to (3) and $(Ax12)$, we get the desired $\Gamma \vdash_{PCON} \neg (\varphi \to \psi)$.

- $\neg \to I_2$: Similar.

- $\neg \to E$: The premises of the rule are $\neg(\varphi \to \psi)$, and two sub-derivations of χ from the assumption-rules $\Gamma, \varphi \vdash_{N_{PCON}} \neg \psi$ and $\Gamma, \neg \varphi \vdash_{N_{PCON}} \psi$. By the induction hypothesis, (1) $\Gamma \vdash_{PCON} \neg(\varphi \to \psi)$, (2) $\Gamma \vdash_{PCON} (\varphi \to \neg \psi) \to \chi$ and (3) $\Gamma \vdash_{PCON} (\neg \varphi \to \psi) \to \chi$. By applying (MP) to (1) and $(Ax12)$, we get (4) $\Gamma \vdash_{PCON} (\varphi \to \neg \psi) \vee (\neg \varphi \to \psi)$. By applying (MP) twice to $(Ax8)$, (2) and (3), we get (5) $\Gamma \vdash_{PCON} ((\varphi \to \neg \psi) \vee (\neg \varphi \to \psi)) \to \chi$. By applying (MP) to (4) and (5) we get the required $\Gamma \vdash_{PCON} \chi$.

dni, dne: Obvious, by $(Ax9)$.

4.3.7 NL motivation of $PCON$

4.3.7.1 Introduction

In this section I provide a motivation for $PCON$, drawing upon certain uses of conjunction and disjunction in natural language (here[5] – English), uses related to difference in *intonational stress* (in speaking) due to focus. This will establish the $PCON$ axioms as *non-arbitrary*, not merely "blindly" produced by the general method of modifying fal-

---

[5] Similar intonational effects are present also in French; an informer confirmed the same about conjunction in German. As a native speaker of Hebrew, I can confirm the same about Hebrew.

4.3 Extending the scope of connexivity: Poly-connexivity

sification conditions, but correspond to a certain abstraction of coordinated sentences in natural language, yet different from the abstraction on which Classical logic is based.

For an explanation of the linguistics nomenclature used in the section, the reader unfamiliar with it may consult the glossary in Chapter 9.

The abstraction is based on a difference of the *intonational stress* pattern in natural language sentences (in speech), in particular in dialogues.

I do not aim toward a comprehensive analysis of the effects of intonational stress on meaning, a rich topic transcending by far the scope of this book. I just review certain aspects of the impact of intonational stress on coordination, to an extent sufficient to make the connection with $PCON$'s axioms for conjunction and disjunction. Typically, such differences in intonational stress are associated with *focus*, where the focused part has *alternatives*. See Rooth [102] for a detailed discussion.

In particular, I will not discuss the details of the "mechanism" by which a difference in intonational stress brings about specific differences in meaning. This "mechanism" involves a complex analysis of the relationship of intonational stress and sets of *alternatives* generated by such a stress, an analysis beyond the scope of this book. All I want is to point out the existence of such a connection, rendering the connexive axioms of $PCON$ non-arbitrary.

Another appeal to NL phenomena to motivate contra-classicality can be found in Omori and Wansing [87], where *negative concord* is used to motivate *demi-negation*, giving rise to a non-classical double negation, as found in Kamide's extensions of Nelson's **N4** logic.

The following notational convention is adhered to throughout this section: for an expression 'e', '$\langle e \rangle_F$' ('F' for 'focus') denotes 'e' with an intonational stress.

In the following section I review some linguistic observations about the interaction of conjunction and disjunction with intonational stress.

Then, in Section 4.3.7.5 I relate those observations to the connectives of the Poly-Connexive logic $PCON$ and their characterizing axioms.

For the benefit of a reader with no linguistics background and for self-containment, there is a glossary in Chapter 9 explaining and exemplifying the linguistics terminology employed (mainly in Section 4.3.7.2).

4.3.7.2 Coordination particles and intonational stress

A hallmark of logical object-languages is that they are *freely generated* from some set of atomic formulas. Thus, for *any* formulas φ and ψ, their conjunction $\varphi \wedge \psi$ and disjunction $\varphi \vee \psi$ are well-formed formulas whenever those connectives are present.

While sentential combination via connectives is also present in (some) natural languages, however, as typically used, no unrelated sentences are ever combined in making assertions. Natural languages have a richer structure allowing also for *sub-sentential coordination*, either as *constituent coordination* or as *non-constituent coordination*, as in

$$\begin{array}{l} \text{She sings and/or dances} \\ \text{She loves Bill and/or John} \\ \text{Bill and/or John love her} \end{array} \quad (4.3.25)$$

It is important to realize that because $PCON$ is a propositional Connexive logic, I am not concerned here with sentences with quantified subject and/or object, like

everyone/every girl/someone/some girl loves Bill and/or John

involving issues of conjunction reduction, the latter not always preserving semantic equivalence. While

Everyone sings and dances ≡ Everyone sings and everyone dances
$$(4.3.26)$$

4.3 Extending the scope of connexivity: Poly-connexivity

we have

$$\text{Everyone sings or dances} \neq \text{Everyone sings or everyone dances} \quad (4.3.27)$$

For the simple, non-quantified subjects and objects used here, the semantic equivalence preservation by a translation to sentential coordination is justified.

Thus, the sentences in (4.3.25) are semantically equivalent to their respective expansion to sentential coordinations.

$$\begin{array}{l} \text{She sings and/or she dances} \\ \text{She loves Bill and/or she loves John} \\ \text{Bill loves her and/or John loves her} \end{array} \quad (4.3.28)$$

To avoid complex theories of possible connections between conjuncts and disjuncts, I will take coordinated sentences resulting from translation of sentences with sub-sentential coordination as indicating the semantic connection between the coordinated subsentence; the connection arising from *sharing* a sub-sentential phrase.

The following example, based on Szabolcsi and Haddican [115], can be incorporated into this framework, where the 'natural pairing' considered in Szabolcsi and Haddican [115] is explicated by constituent sharing.

Example 4.3.22 (enrollment). *Imagine the following scenario. The enrollment regulations in a certain college are such that a student is required to enroll into two courses, both belonging to one of two groups. Science, consisting of Physics and Mathematics, and Humanities, consisting of Art and Music. The courses in each group constitute the natural pairings. The following dialogue takes place between participants*

A (a student) and B (a secretary).

$$\mathcal{D}_{c,1} :: \quad \begin{array}{l} A: \text{I want to enroll in Mathematics and in Art} \\ \\ B: \begin{array}{l} \text{No!} \\ \text{You can enroll in Mathematics and not in Art} \\ \text{(but in Physics),} \\ \text{or in Art and not in Mathematics (but in Music)} \end{array} \end{array}$$

(4.3.29)

The student's request violates the natural pairing between conjuncts, breaking the enrollment regulations. The secretary's correction restores the natural pairing by directing the student to adhere to the enrollment regulations.

As will be shown below, this kind of meaning connection gives rise to differences in intonational stress that lead to the connexive axioms.

4.3.7.3 Conjunction and intonational stress

The following observations are from Gacić [39].

A sentence like (4.3.30) below can have two readings.

$$\text{He did not visit Colombia and Brazil.} \quad (4.3.30)$$

- It is possible to interpret (4.3.30) as (4.3.31) below

    He did not visit both countries but only one of the two. (4.3.31)

    where it is not the case that he visited *both* Colombia and Brazil, so he either visited Colombia (and not Brazil) or Brazil (and not Colombia).

4.3 Extending the scope of connexivity: Poly-connexivity

- But it is also possible to interpret (4.3.30) as (4.3.32)

> He visited neither country. (4.3.32)

where he didn't visit either of the two countries, i.e. he didn't visit Colombia and he didn't visit Brazil.

The two readings correlate with different *intonational stress patterns* of the sentence (4.3.30). Stress on the connective **and**, as in (4.3.33), is normally required for the reading in (4.3.31), whereas the whole conjunction is stressed, as in (4.3.34), for the reading in (4.3.32).

> He ⟨did not⟩$_F$ visit Colombia ⟨and⟩$_F$ Brazil (4.3.33)

> He ⟨did not⟩$_F$ visit ⟨Colombia and Brazil⟩$_F$ (4.3.34)

Negation is stressed as well in both patterns (4.3.33, 4.3.34).

There is a dialogue-related phenomenon that supports the correspondence with intonational stress variation observed above. The two intonational stress patterns are compatible with *different continuations* of (4.3.30). Namely, when the stress is placed on the connective, the sentence is naturally followed by either an assertion of one (4.3.36 a) or the other (4.3.36 b) country having been visited. However, it is not felicitous to bring up as a follow-up a third, so far unmentioned, third country as in (4.3.37).

> He did not visit Colombia ⟨and⟩$_F$ Brazil (4.3.35)

✓ a. He visited (only/just) Colombia.
✓ b. He visited (only/just) Brazil.
✓ c. He visited either Colombia or Brazil, but I'm not sure which one he picked in the end. (4.3.36)

> #He visited Peru. (4.3.37)

Conversely, when the whole conjunction is stressed as in (4.3.38), asserting that he visited only Colombia (4.3.39 a) or only Brazil (4.3.39

b) is an infelicitous continuation, whereas a continuation containing an alternative that is not found in either of the conjuncts (Peru in (4.3.40)) is now compatible with the initial utterance (4.3.38).

$$\text{He did not visit } \langle \text{Colombia and Brazil} \rangle_F \quad (4.3.38)$$

$$\begin{array}{l} \# \, a.\text{He visited (only/just) Colombia.} \\ \# \, b.\text{He visited (only/just) Brazil.} \end{array} \quad (4.3.39)$$

$$\checkmark \text{He visited Peru.} \quad (4.3.40)$$

As observed in Han and Romero [44], there is another dialogue-related corroboration of the distinction between the meanings originating from different intonational stress[6], namely, *question answering*: polar (y/n) questions vs. alternative (alt) questions.

- An intonational stress on **and** is a felicitous answer to an alt-question. Thus, (4.3.33) forms a felicitous answer to the question

   $$\text{Which Latin American countries did he not visit?} \quad (4.3.41)$$

- On the other hand, an intonational stress on the whole coordinated VP is a felicitous answer to a y/n-question. Thus, (4.3.46) forms a felicitous (negative) answer to the question

   $$\text{Did he visit Colombia and Brazil?} \quad (4.3.42)$$

4.3.7.4 Disjunction and intonational stress

A similar[7] disambiguation by intonational stress can be observed also for disjunction, although less frequently mentioned in the literature. See, for example, the recent Lungu, Fălăus, and Panzeri [65].

---

[6]The observation in Han and Romero [44] relates to a yet another pattern of intonational stress in coordinated sentences.

[7]There is another stress pattern associated with disjunction, studied in Han and Romero[44].

4.3 Extending the scope of connexivity: Poly-connexivity

A sentence like (4.3.43) below can also have two readings.

$$\text{He did not visit Colombia or Brazil.} \quad (4.3.43)$$

- One reading of (4.3.43) is

$$\text{He did not visit Colombia and he did not visit Brazil} \quad (4.3.44)$$

- Another reading of (4.3.43) is

$$\text{It is one of Colombia or Brazil he did not visit} \quad (4.3.45)$$

Again, the two readings correlate with different *intonation stress patterns* of the sentence (4.3.43). Stress on the connective or, as in (4.3.46), is normally required for the reading in (4.3.45), whereas the whole disjunction is stressed, as in (4.3.47), for the reading in (4.3.44).

$$\text{He } \langle\text{did not}\rangle_F \text{ visit Colombia } \langle\text{or}\rangle_F \text{ Brazil} \quad (4.3.46)$$

$$\text{He } \langle\text{did not}\rangle_F \text{ visit } \langle\text{Colombia or Brazil}\rangle_F \quad (4.3.47)$$

Again, negation is stressed as well in both patterns (4.3.46, 4.3.47).

In terms of continuations (in a dialogue) we have the following.

- Under the reading (4.3.44), (4.3.43) can have (4.3.40) as a felicitous continuation.

- Under the reading (4.3.45), (4.3.43) can have

$$\begin{array}{l} \checkmark a. \text{ He did not visit Colombia.} \\ \checkmark a. \text{ He did not visit Brazil.} \end{array} \quad (4.3.48)$$

as a felicitous continuation, but not (4.3.40).

A similar relation to answering questions is present also in the case of disjunction.

- An intonational stress on or is a felicitous answer to an alt-question. Thus, (4.3.46) is a felicitous answer to the alt-question (4.3.41).

- On the other hand, an intonational stress on the whole coordinated VP is a felicitous answer to a y/n-question. Thus, (4.3.46) forms a felicitous (negative) answer to the question (4.3.42).

4.3.7.5 $PCON$ and intonational stress

An observant reader will have noticed by now that the $PCON$ axioms **Ax10** and **Ax11** in Section 4.3.2 correspond to meanings of natural language coordinated sentences resulting from constituent coordination, the latter *expressed by an intentionally stressed coordination marker*.

- The axiom **Ax10** corresponds to the stressed conjunction 'and', as in (4.3.33), where the negation negates that *both* of the conjuncts expressed by the stressed conjunction particle obtain, but only one of them does.

- Similarly, the axiom **Ax11** corresponds to the stressed disjunction 'or', as in (4.3.46).

Thus, the major parameter in the above analysis seems to be whether an intonational stress is placed on the connectives and/or, as in (4.3.33) and (4.3.46), or on the whole conjunction/disjunction as a coordinated constituent, as in (4.3.34) and (4.3.47).

The portrayed situation emerging is that $PCON$ is not an *arbitrary logic*: it is a logic of connectives with an *intonational stress* in natural language coordinated sentences corresponding to sentences with constituent coordination.

This analysis adheres to the two observations about connexivity mentioned above.

1. The two arguments of a (binary) connective are mutually related when the connective is read connexively. The relatedness manifests itself explicitly in the constituent coordination in the sentences giving rise to the sentential coordination; this expands the observation about relatedness of the antecedent and consequent in a valid connexive conditional.

2. The connexivity emerges when the coordination is *negated*, adhering to the observation of falsification conditions giving rise to connexivity.

4.3.7.6 Negating the connexive conditional and intonational stress

Once a connection between connectives and intonational stress has been established, I utilize this connection and apply it also to the orthodox connexive conditional. I consider again the two ways of negating the conditional:

$$(r) \; \neg(\varphi \rightarrow \psi) \equiv \varphi \rightarrow \neg\psi$$

$$(l) \; \neg(\varphi \rightarrow \psi) \equiv \neg\varphi \rightarrow \psi$$

(4.3.49)

Recall that negating according to (r) has been proposed by Wansing [119, (axiom a5)], while negating according to (l) was proposed in Francez [33], and studied further in Omori [84].

Consider first the following dialogue, where φ and ψ are regimentations of two NL sentences,

$$\mathcal{D}_{cond,l} :: \begin{array}{l} A: \text{ does } \langle\varphi\rangle_\mathsf{F} \text{ imply } \psi? \\ B: \text{ No! } \langle\neg\varphi\rangle_\mathsf{F} \text{ implies } \psi. \end{array}$$

(4.3.50)

In A's question, the antecedent of the conditional is with an intonational stress and is in focus. Here the only alternative to φ (see 'intonational stress' in the glossary in Chapter 9 for focal alternatives) is $\neg\varphi$. The 'No' in B's response denies the implication $\varphi \rightarrow \psi$, expressing the negation as

$\neg\varphi\to\psi$ with the *same* intonational stress pattern. This embodies pattern (*l*) above of negating the conditional.

A similar effect of intonational stress can be exhibited also without questions.

$$\mathcal{D}'_{cond,l} :: \begin{array}{l} A: \langle\varphi\rangle_F \text{ implies } \psi. \\ B: \text{No! } \langle\neg\varphi\rangle_F \text{ implies } \psi. \end{array} \qquad (4.3.51)$$

Example 4.3.23. *Suppose A and B are two fans of a football team X, arguing about the influence of the weather on the result of a match involving X.*

$$\begin{array}{l} A: \text{if } \langle\text{it rains}\rangle_F, \text{ team X will win.} \\ B: \text{No! if } \langle\text{it does not rain}\rangle_F, \text{ team X will win.} \end{array} \qquad (4.3.52)$$

Next, consider the following dialogue:

$$\mathcal{D}_{cond,r} :: \begin{array}{l} A: \text{does } \varphi \text{ imply } \langle\psi\rangle_F? \\ B: \text{No! } \varphi \text{ implies } \langle\neg\psi\rangle_F. \end{array} \qquad (4.3.53)$$

Here the consequent is with intonational stress, and the emerging negation pattern is (*r*).

$$\begin{array}{l} A: \text{if it rains, } \langle\text{team X will win}\rangle_F. \\ B: \text{No! if it it rains, } \langle\text{team Y will win}\rangle_F. \end{array} \qquad (4.3.54)$$

Example 4.3.24 (continued). *Suppose the other team in the match is Y, an alternative winner, so the implication is that X does not win.*

By proposing this formal correspondence, I justified the *programmatic* point in introducing $PCON$, namely, *extending the scope of connexivity*, currently confined to the interaction of negation and the conditional, to an interaction of negation with other binary connectives, requiring a *connection in content* between their two arguments. I have shown that the difference between the $PCON$ axioms and the classical logic corresponding axioms captures a difference in meaning of NL sentences, depending in different focal stress patterns they employ.

After all, this dependency on connection in content was the initial motivation for connexive logics, as evident from their name.

Chapter 5

Connexivity and relevance

5.1 Introduction

Connexive logics share a common source with the "cousin" family of logics, *Relevance logics* (known also as Relevant logics), the latter originating from Anderson and Belnap [1], already mentioned in Section 1.2. Both families of logics share the view that in a valid conditional proposition there should be a connection between the antecedent and the consequent *in terms of content* (and not just truth-value). This view goes as far as the following statement by Routley [103] (p. 393):

> [t]he general classes of connexive and relevant logics are one and the same.

Thus, it comes as no surprise that there were attempts to combine connexivity and relevance in one and the same logic. The idea of combining Relevance logics with Connexive logics is discussed in the Introduction section in Wansing [122].

In this Chapter, I combine two of the major revolts against the material conditional as adequately capturing the use of the conditional in everyday (non-mathematical) reasoning, namely *Relevance logic(s)* and *Connexive logic(s)*. The presentation is based on Francez [34].

5.2 Connexive Relevant logic

5.2.1 Introduction

I present a connexive extension $\mathcal{L}_{rc}$, a logic of relevance *and* connexivity, of the one of the most typical Relevance logics, namely $\mathbf{R}_\to$, the implicational fragment of Anderson and Belnap's $\mathbf{R}$ [1]. Such a combination should produce a logic endowed with a conditional reflecting both aspects of content that underly the two combined logics.

As it turns out, observed by Weiss [126], the implication-negation fragment $C_{\to,\neg}$ of Wansing's Connexive logic C is an expansion of $\mathcal{L}_{rc}$ with a weakening axiom $\varphi \to (\psi \to \varphi)$. Naturally, $\mathcal{L}_{rc}$, being a Relevance logic in addition to being a Connexive logic, has to abandon Weakening. See Weiss [126] for a general discussion of those (and other) logics in a common framework.

My definitional tool is a natural-deduction (ND) proof system $\mathcal{N}_{rc}$ (presented in Section 5.2.2.1), combining two methods adopted from the respective underlying families of logic:

- Providing rules for negating the conditional, adopted from the methodology of defining connexive logics, avoiding uniform I/E-rules for negation.

- Keeping track of the "*use*" of an assumption in a derivation as a necessary condition for its discharge by a rule application, adopted

5.2 Connexive Relevant logic 131

from the methodology of defining Relevance logics, avoiding vacuous discharge.

In [74] McCall presents such a system, adopting the 'use tracing' technique of the ND-system of Anderson&Belnap [1], not specifically associated with **R**$_\rightarrow$.

Choosing ND as the definitional tool has the advantage of focusing on proofs from assumptions (derivations), and not merely on formal theorems (theses), as is the tendency in axiomatic, Hilbert-like presentations. See Schroeder-Heister [109] for a general discussion of this difference, and Avron [5] for a discussion of this difference in the context of Relevance logics.

The ND-system $\mathcal{N}_{rc}$ is shown to be deductively equivalent to an axiomatic (Hilbert-like) presentation $\mathcal{H}_{rc}$. No model theory for $\mathcal{L}_{rc}$ is presented here. A model-theory sound and complete w.r.t. $\mathcal{H}_{rc}$ can be found in Weiss [126]. Also, Weiss uses his model-theory to establish that $\mathcal{H}_{rc}$ (and therefore also $\mathcal{L}_{rc}$) satisfies the variable-sharing condition.

5.2.2 The natural-deduction system $\mathcal{N}_{rc}$

I consider here an implication-negation fragment. I use '$\rightarrow$' for the conditional and '$\neg$' for negation.

5.2.2.1 Defining $\mathcal{N}_{rc}$

The introduction and elimination rules (I/E rules) are presented in Figure 5.1. As usual, an assumption enclosed in square brackets indicates a discharged assumption. Its index (discharge index) appears as a superscript on the instance of the rule the application of which discharges this

$$\frac{[\varphi]_i}{\varphi_i} \, (Ass), \; i \text{ fresh} \qquad (5.2.1)$$

$$\frac{\begin{array}{c}[\varphi]_i \\ \vdots \\ \psi_{\alpha \cup \{i\}}\end{array}}{(\varphi \to \psi)_\alpha} \, (\to^+ I^i) \qquad \frac{(\varphi \to \psi)_\alpha \quad \varphi_\beta}{\psi_{\alpha \cup \beta}} \, (\to^+ E) \qquad (5.2.2)$$

$$\frac{\begin{array}{c}[\varphi]_i \\ \vdots \\ (\neg \psi)_{\alpha \cup \{i\}}\end{array}}{(\neg(\varphi \to \psi))_\alpha} \, (\to^- I^i) \qquad \frac{(\neg(\varphi \to \psi))_\alpha \quad \varphi_\beta}{(\neg \psi)_{\alpha \cup \beta}} \, (\to^- E) \qquad (5.2.3)$$

$$\frac{(\neg\neg\varphi)_\alpha}{\varphi_\alpha} \, (dne) \qquad \frac{\varphi_\alpha}{(\neg\neg\varphi)_\alpha} \, (dni) \qquad (5.2.4)$$

Figure 5.1: The $\mathcal{N}_{rc}$ rules

assumption. A characteristic feature of $\mathcal{N}_{rc}$ is that it has separate I/E-rules for positive occurrences of the conditional and for negative (i.e., negated) occurrences. Note that the non-vacuous discharge restriction applies both to the positive and negative I-rules for the conditional.

Derivations (tree-shaped, Prawitz's style), ranged over by $\mathcal{D}$, are defined inductively as usual, iterating rule applications starting from assumptions. Any interim occurrence of a formula φ in a derivation has a subscript α tracking the (open) assumptions this occurrence of φ has used, on which this occurrence of φ depends. See Avron [5] for a critical discussion of 'use' in the literature of Relevance logics.

Definition 5.2.8 (use of assumptions). *An assumption, say an instance of φ, is* used *in a derivation $\mathcal{D}$ when that instance of φ serves as a premise for the application of some rule within $\mathcal{D}$.*

5.2 Connexive Relevant logic

Use is *propagated* along a derivation according to the manipulation of indices of used assumption by the various rule applications along the derivation.

Notation: When the set α is a singleton the set brackets are omitted, and when α is empty the subscript itself is omitted.

By convention, I assume that in the $\mathcal{N}_{rc}$-derivations the open assumptions Γ are subscripted by $1\cdots n$ for some $n \geq 0$, and let $\hat{n} = \{1, \cdots, n\}$.

Remarks about the rules of $\mathcal{N}_{rc}$:

ad (*Ass*): A similar way of formulating the introduction of an assumption is attributed by von Plato [94] to Gentzen. The rule was intended to make less awkward the derivation of $\varphi \supset \varphi$, and was later abandoned by Gentzen.

I revive this formulation here in order to make it explicit that φ, assumed in the premise, is also an explicit conclusion, *carrying the same index*! Thus, when this rule is applied, it uses the assumption in the premise, allowing later discharge of the latter. We thus get

$$\dfrac{\dfrac{[\varphi]_1}{\varphi_1}\,(Ass)}{\varphi \to \varphi}\,(\to^+ I^1) \tag{5.2.5}$$

To avoid notational clutter, I will omit applications of this rule in example derivations in which it has no beneficial contribution as it has in (5.2.5) above.

Regarding the freshness of i, strictly speaking, i is chosen as the *least* natural number not used as an index in the derivation at an earlier stage. I will usually ignore this strictness and just relate to i is being fresh, not used so far.

ad ($\to^+ I$): This is the original I-rule of Anderson and Belnap [1] for introducing the relevant conditional. The assumption of the antecedent is only dischargeable if used during the sub-derivation of the consequent ψ from the assumed antecedent φ. This condition is enforced by requiring ψ to have an index containing i, the fresh index of the assumption.

ad ($\to^+E$): This is a "relevantized" version of modus-ponens.

The conclusion uses the union of the assumption sets used by both premises. Note that this rule, together with ($\to^-E$), are rules in which use is *propagated*, the assumption set index of the conclusion growing.

ad ($\to^-I$): This rule endows the conditional '$\to$' its connexivity, by changing its falsification condition. I have originally introduced this rule in Francez [33], without the tracking of indices needed for relevance, motivating it by certain uses of negated conditionals in natural language.

Note that as long as no additional connectives are included in the object-language, there are three sources of $\neg\psi$, the negated conclusion of this rule:

1. A recursive application of the ($\to^-I$) rule itself.
2. An application of ($\to^-E$).
3. An application of (dni).

ad ($\to^-E$): This E-rule is naturally associated with the corresponding ($\to^-I$) rule, retrieving the grounds for introduction of the negated conditional.

ad (dni), (dne): Note that both rules have no effect on the used assumptions index, merely preserving it.

□

Remark 5.2.9. *In $\mathcal{N}_{rc}$, an assumption $[\varphi]_i$ can be used within ($\to^+I$) or ($\to^-I$), before being discharged, in three ways:*

1. *As a premise of an application of the assumption rule (Ass).*

5.2 Connexive Relevant logic

2. As a minor premie of $(\to^+ E)$ or $(\to^- E)$; in this case, there is for some ξ one of the following two a sub-derivations:

$$\frac{[\varphi]_i \quad (\varphi \to \xi)_\beta}{\xi_{\beta \cup i}} (\to^+ E) \qquad \frac{[\varphi]_i \quad \neg(\varphi \to \xi)_\beta}{\neg \xi_{\beta \cup i}} (\to^- E)$$

with $\beta \subseteq \alpha$ in both cases.

3. As a major premise of $(\to^+ E)$ or $(\to^- E)$; in this case, $\varphi = \chi \to \xi$ or $\varphi = \neg(\chi \to \xi)$ for some χ, ξ and there is one of the following two a sub-derivations:

$$\frac{[\chi \to \xi]_i \quad \chi_\beta}{\xi_{\beta \cup i}} (\to^+ E) \qquad \frac{[\neg(\chi \to \xi)]_i \quad \chi_\beta}{\neg \xi_{\beta \cup i}} (\to^- E)$$

with $\beta \subseteq \alpha$ in both cases.

□

Denote by $\vdash_{\mathcal{N}_{rc}} \Gamma : \varphi$ the derivability (deducibility) in $\mathcal{N}_{rc}$ of φ from the open assumptions Γ. When Γ is empty, φ is a thesis (formal theorem) of $\mathcal{L}_{rc}$.

Note that the rules $(\to^+ I/E$ are the defining rules for $\mathbf{R}_\to$. The arguments for $\mathcal{N}_{rc}$ rendering $\mathcal{L}_{rc}$ a Relevance logic are the same arguments put forward originally by Anderson and Belnap for $\mathbf{R}_\to$ being a Relevance logic due to their use-tracking ND-system; I will not repeat them here.

5.2.2.2 The connexivity of $\mathcal{L}_{rc}$

The examples below establish the connexivity of $\mathcal{L}_{rc}$.

Example 5.2.25 (A_i, $i = 1, 2$).

$$\frac{\dfrac{\dfrac{[\varphi]_1}{\varphi_1}\,(Ass)}{\neg\neg\varphi_1}\,(dni)}{\neg(\varphi\to\neg\varphi)}\,(\to^- I^1) \qquad \frac{\dfrac{[\neg\varphi]_1}{\neg\varphi_1}\,(Ass)}{\neg(\neg\varphi\to\varphi)}\,(\to^- I^1) \tag{5.2.6}$$

It is interesting to observe that Orlov (as reported by Došen in [16], discussed also in Restall [101]), not aiming at connexivity, had a weaker version of this thesis as an axiom, fitting his conception of negation in Relevance logic:

$$\varphi\to\neg(\varphi\to\neg\varphi) \tag{5.2.7}$$

a contraposition of a variant of Reductio:

$$(\varphi\to\neg\varphi)\to\neg\varphi \tag{5.2.8}$$

The other axiom added by Orlov for a characterization of Relevance logic negation is contraposition:

$$(\varphi\to\neg\psi)\to(\psi\to\neg\varphi) \tag{5.2.9}$$

Example 5.2.26 (B_i, $i = 1, 2$).

$$\frac{\dfrac{\dfrac{\dfrac{[\varphi\to\psi]_1\quad[\varphi]_2}{\psi_{\{1,2\}}}\,(\to^+ E)}{\neg\neg\psi_{\{1,2\}}}\,(dni)}{\neg(\varphi\to\neg\psi)_1}\,(\to^- I^2)}{(\varphi\to\psi)\to\neg(\varphi\to\neg\psi)}\,(\to^+ I^1) \qquad \frac{\dfrac{\dfrac{[\varphi\to\neg\psi]_1\quad[\varphi]_2}{\neg\psi_{\{1,2\}}}\,(\to^+ E)}{\neg(\varphi\to\psi)_1}\,(\to^- I^2)}{(\varphi\to\neg\psi)\to\neg(\varphi\to\psi)}\,(\to^+ I^1) \tag{5.2.10}$$

As for ($asym$), namely $\nvdash_{\mathcal{N}_{rc}} (\varphi\to\psi)\to(\psi\to\varphi)$, this is established in Weiss [126] by constructing a counter model in his model-theory.

5.2.2.3 Negation inconsistency of $\mathcal{L}_{rc}$

I now show that, like many other Connexive logics, $\mathcal{L}_{rc}$ is negation-inconsistent.

5.2 Connexive Relevant logic

Proposition 5.2.29 (negation-inconsistency of $\mathcal{L}_{rc}$). *$\mathcal{L}_{rc}$ is negation-inconsistent.*

Proof. I show that[1]

$$\vdash_{\mathcal{N}_{rc}} (((\neg\varphi\to\varphi)\to\neg\varphi)\to\neg((\neg\varphi\to\varphi)\to\neg\varphi))$$

The result follows, since this is a contradiction to A_1.

The derivation is as follows.

$$\cfrac{\cfrac{\cfrac{\cfrac{\cfrac{[((\neg\varphi\to\varphi)\to\neg\varphi)]_1 \quad [\neg\varphi\to\varphi]_2}{\neg\varphi_{\{1,2\}}} \quad [\neg\varphi\to\varphi]_2}{\varphi_{\{1,2\}}}(\to^+E)}{\neg\neg\varphi_{\{1,2\}}}(dni)}{\neg((\neg\varphi\to\varphi)\to\neg\varphi)_{\{1\}}}(\to^-I^2)}{(((\neg\varphi\to\varphi)\to\neg\varphi)\to\neg((\neg\varphi\to\varphi)\to\neg\varphi))}(\to^+I^1)$$

$\square$

Clearly, $\mathcal{L}_{rc}$, in spite of being negation inconsistent, is not trivial.

5.2.2.4 More examples

Example 5.2.27. *Below is a proof of one of the axioms of* $\mathbf{R}_\to$ *(see Section 5.3).*

$$\vdash_{\mathcal{N}_{rc}} (\varphi\to\psi)\to((\chi\to\varphi)\to(\chi\to\psi)) \tag{5.2.11}$$

---

[1]This formula was identified as contradicting A_1 already by McCall in Section 29.8 of Anderson and Belnap [1] (p. 436).

The derivation is

$$\cfrac{\cfrac{\cfrac{\cfrac{\cfrac{[\chi\to\varphi]_2 \quad [\chi]_1}{\varphi_{\{1,2\}}}(\to^+E) \quad [\varphi\to\psi]_3}{\psi_{\{1,2,3\}}}(\to^+E)}{\chi\to\psi_{\{2,3\}}}(\to I^1)}{(\chi\to\varphi)\to(\chi\to\psi)_3}(\to^+I^2)}{(\varphi\to\psi)\to((\chi\to\varphi)\to(\chi\to\psi))}(\to^+I^3)$$

It easy to see that no assumption was vacuously discharged.

Example 5.2.28. *Here is an example of a derivation of a non-axiom formal theorem of* $\mathbf{R}_\to$ *(due to Mordechai Wajsberg) in* $\mathcal{N}_{rc}$.

$$\vdash_{\mathcal{N}_{rc}} ((\varphi\to\varphi)\to\psi)\to\psi \tag{5.2.12}$$

The derivation is

$$\cfrac{\cfrac{[(\varphi\to\varphi)\to\psi]_1 \quad \cfrac{\cfrac{[\varphi]_2}{\varphi_2}(Ass)}{\varphi\to\varphi}(\to^+I^2)}{\psi_1}(\to^+E)}{((\varphi\to\varphi)\to\psi)\to\psi}(\to^+I^1)$$

Example 5.2.29.

$$\vdash_{\mathcal{N}_{rc}} \neg(\varphi\to(\psi\to\chi)) : \varphi\to(\psi\to\neg\chi) \tag{5.2.13}$$

The derivation is:

$$\cfrac{\cfrac{\cfrac{\cfrac{\cfrac{\neg(\varphi\to(\psi\to\chi))_1 \quad [\varphi]_2}{\neg(\psi\to\chi)_{\{1,2\}}}(\to^-E) \quad [\psi]_3}{\neg\chi_{\{1,2,3\}}}(\to^-E)}{(\psi\to\neg\chi))_{\{1,2\}}}(\to^+I^3)}{(\varphi\to(\psi\to\neg\chi))_1}(\to^+I^2)$$

This derivation naturally generalizes to show

$$\vdash_{\mathcal{N}_{rc}} \neg(\varphi_1\to(\cdots\to\varphi_n)\cdots) : (\varphi_1\to(\cdots\to\neg\varphi_n)\cdots), \; n\geq 3 \tag{5.2.14}$$

5.2 Connexive Relevant logic

Example 5.2.30.

$$\vdash_{\mathcal{N}_{rc}} \varphi \to \psi, \neg(\psi \to \chi) : \neg(\varphi \to \chi) \tag{5.2.15}$$

the derivation is:

$$\cfrac{\neg(\psi \to \chi)_2 \quad \cfrac{(\varphi \to \psi)_1 \quad [\varphi]_3}{\psi_{\{1,3\}}} (\to^+ E)}{\cfrac{\neg \chi_{\{1,2,3\}}}{\neg(\varphi \to \chi)_{\{1,2\}}} (\to^- I^3)} (\to^- E)$$

Proposition 5.2.30 (conservativity). $\mathcal{N}_{rc}$ *is conservative over* $\mathbf{R}_\to$.

Proof: Suppose $\vdash_{\mathcal{N}_{rc}} \Gamma : \chi$ where Γ, χ are negation free. The proof is by induction on the last $\mathcal{N}_{rc}$-rule applied in the derivation. For avoiding notational clutter, I ignore the use-tracking indices. Only rules which have a negation free conclusion and are not rules of $\mathbf{R}_\to$ need to be considered. This leaves only (dne). Consider now how the premise of (dne), namely $\neg\neg\chi$ was derived. Note that since Γ contains only $\neg$-free assumptions, the only way negation enters the derivation is by means of an (dni) application. Call the point where this happens the negation injection point.

There are two possibilities.

1. Via $(\to^- E)$: That is, ψ is $\neg\chi$, and the $(\to_- E)$-application looks like

$$\cfrac{\cfrac{\neg(\varphi \to \neg\chi) \quad \varphi}{\neg\neg\chi} (\to^- E)}{\chi} (dne) \tag{5.2.16}$$

   By the induction hypothesis, the minor premise φ is derivable in $\mathbf{R}_\to$. Assume w.l.o.g. that the major premise was introduced by the following sub-derivation, containing the negation injection point.

$$\frac{\dfrac{[\varphi]_i}{\mathcal{D}} }{\dfrac{\chi}{\neg(\varphi\to\neg\chi)}\;(dni)}\;(\to^- I^i) \qquad (5.2.17)$$

But then, (5.2.17) can be replaced by

$$\dfrac{\begin{array}{c}[\varphi]_i\\ \mathcal{D}\\ \chi\end{array}}{\varphi\to\chi}\;(\to^+ I^i) \qquad (5.2.18)$$

and (5.2.16) cab be replaced by

$$\dfrac{\varphi\to\chi \quad \varphi}{\chi}\;(\to +^+ E) \qquad (5.2.19)$$

an $\mathbf{R}_\to$-derivation, as required.

2. Via (dni): immediate.

5.2.3 On the negated conditional

So, how does the connexive-relevant negated conditional differ from its counterpart in $\mathbf{R}_{\to,\neg}$? As a representative of the latter, I take the following two $(\neg I)$ and $(\neg E)$ rules from Mares [67] (Section 7.3). One might use an explicit contradiction to avoid an appeal to $\bot$, which is not used in $\mathcal{N}_{rc}$. These are general rules for $\neg\varphi$, independent on the form of φ.

$$\dfrac{\begin{array}{c}[\varphi]_i\\ \vdots\\ \bot_{\alpha\cup i}\end{array}}{(\neg\varphi)_\alpha}\;(\neg I^i),\;i\text{ fresh} \qquad \dfrac{\varphi_\alpha \quad \neg\varphi_\beta}{\bot_{\alpha\cup\beta}}\;(\neg E) \qquad (5.2.20)$$

Note the absence of $(\bot E)$ in order not to validate explosion.

5.2 Connexive Relevant logic

When applied to derive a negated conditional, we get

$$\frac{\begin{array}{c}[\varphi\to\psi]_i\\ \vdots\\ \bot_{\alpha\cup i}\end{array}}{(\neg(\varphi\to\psi))_\alpha}\,(\neg I^i) \qquad \frac{(\varphi\to\psi)_\alpha \quad \neg(\varphi\to\psi)_\beta}{\bot_{\alpha\cup\beta}}\,(\neg E) \qquad (5.2.21)$$

So, a negated conditional means that assuming the (relevant) conditional leads to a contradiction. Thus, there is *no* proof of ψ from an assumption φ using that assumption. This absence of proof may originate from several reasons; in particular, it leaves open the possibility that ψ is just *not relevant* to φ, and, hence, not relevantly provable from it.

On the other hand, the connexive-relevant rule $(\neg\to I)$ requires more! It requires that $\neg\psi$, hence ψ itself, *is* relevant to φ, as it requires a proof of ψ from an assumption $\neg\varphi$ *using the latter*!

Thus, the connexive-relevant conditional has a stronger content connection between the antecedent and a consequent of a valid conditional than its relevant counterpart.

5.2.4 A natural deduction-theorem for $\mathcal{L}_{rc}$

The deduction meta-theorem (DT) is usually formulated for axiomatically defined logics. To recapitulate, let $\mathcal{H}$ be some generic axiomatic system over an object-language containing a generic conditional '$\Rightarrow$'.

$$(DT)\ \vdash_\mathcal{H}\Gamma,\varphi:\psi \text{ iff } \vdash_\mathcal{H}\Gamma:\varphi\Rightarrow\psi \qquad (5.2.22)$$

According to Avron [4], the satisfaction of the condition of the deduction theorem, but formulated in terms of a logical consequence relation, constitutes the definition of a binary connective being a conditional.

I will attend to the deduction-theorem for $\mathcal{L}_{rc}$ after the axiomatic definition of the latter, $\mathcal{H}_{rc}$, is presented in Section 5.3 and shown deductively equivalent to $\mathcal{N}_{rc}$.

However, one can conceive a version of the deduction-theorem applicable to natural-deduction systems instead of axiomatic ones. I refer to such version as a natural deduction-theorem (NDT). Such an (NDT) theorem is usually much easier to prove than its axiomatic counterpart because of the way conditionals of various kinds are introduced by natural-deduction I-rules.

Let $\mathcal{N}$ be some generic natural-deduction system over an object-language containing a generic conditional '$\Rightarrow$'.

$$(NDT) \vdash_\mathcal{N} \Gamma, \varphi : \psi \text{ iff } \vdash_\mathcal{N} \Gamma : \varphi \Rightarrow \psi \qquad (5.2.23)$$

The formulation looks identical to the usual formulation of the deduction-theorem, but it pertains to ND-derivations and not to axiomatic derivations (from open assumptions).

I now formulate (NDT_{rc}), the *relevant-connexive* natural deduction-theorem for $\mathcal{N}_{rc}$. The formulation adapts the general case to the presence of assumption tracking indices in $\mathcal{N}_{rc}$.

Theorem 5.2.2 (relevant-connexive natural deduction-theorem for $\mathcal{N}_{rc}$). *For any Γ, φ and ψ:*

positive NDT_{rc}

$$(NDT_{rc}^+) \; i \notin \hat{n} \text{ and } \vdash_{\mathcal{N}_{rc}} \Gamma, \varphi_i : \psi_{\hat{n} \cup i} \text{ iff } \vdash_{\mathcal{N}_{rc}} \Gamma : (\varphi \rightarrow \psi)_{\hat{n}} \qquad (5.2.24)$$

negative NDT_{rc}

$$(NDT_{rc}^-) \; i \notin \hat{n} \text{ and } \vdash_{\mathcal{N}_{rc}} \Gamma, \varphi_i : \neg\psi_{\hat{n} \cup i} \text{ iff } \vdash_{\mathcal{N}_{rc}} \Gamma : \neg(\varphi \rightarrow \psi)_{\hat{n}} \qquad (5.2.25)$$

Proof. **positive NDT**

    only if: Assuming $i \notin \hat{n}$ and $\vdash_{\mathcal{N}_{rc}} \Gamma, \varphi_i : \psi_{\hat{n} \cup i}$, just use ($\rightarrow^+ I$).

if: Assuming $\vdash_{\mathcal{N}_{rc}}(\varphi\to\psi)_{\hat{n}}$, we have

$$\dfrac{\Gamma:(\varphi\to\psi)_{\hat{n}} \quad \dfrac{[\varphi]_i \; (ass)}{\varphi_i}}{\Gamma,\varphi_i:\psi_{\hat{n}\cup i}}(\to^+E)$$

Also, $i \notin \hat{n}$ holds by the freshness of i.

negative NDT

only if: Assuming $i \notin \hat{n}$ and $\vdash_{\mathcal{N}_{rc}}\Gamma,\varphi_i:\neg\psi_{\hat{n}\cup i}$, just use $(\to^-I)$.

if: Assuming $\vdash_{\mathcal{N}_{rc}}\neg(\varphi\to\psi)_{\hat{n}}$, we have

$$\dfrac{\Gamma:\neg(\varphi\to\psi)_{\hat{n}} \quad \dfrac{[\neg\varphi]_i \; (ass)}{(\neg\varphi)_i}}{\Gamma,\varphi_i:(\neg\psi)_{\hat{n}\cup i}}(\to^-E)$$

Again, $i \notin \hat{n}$ holds by the freshness of i.

$\square$

5.3 Axiomatic definition of $\mathcal{L}_{rc}$

5.3.1 Defining $\mathcal{H}_{rc}$

In this section, an axiomatic definition of $\mathcal{L}_{rc}$ by means of a Hilbert-system $\mathcal{H}_{rc}$ is presented and $\mathcal{H}_{rc}$ is shown to be deductively equivalent to $\mathcal{N}_{rc}$. Note that while in large parts of the literature axiomatic presentation are taken as a *definitional*, self-justifying presentation, I consider it as justified by the deductive equivalence to $\mathcal{N}_{rc}$, the latter being the definitional tool adopted.

The development of the axiomatization is similar to that of Wansing's axiomatic definition of the connexive logic C [119], but instead of starting with the axioms of positive (propositional) intuitionistic logic, I start with the (positive) axioms of **R**$_\to$, the positive implicative fragment of **R**, to which negative axioms, inducing connexivity, are added.

Definition 5.3.9 (Axiomatic definition of $\mathcal{L}_{rc}$). *The Hilbert-style axiomatic definition $\mathcal{H}_{rc}$ of $\mathcal{L}_{rc}$ is given by the following axiom schemes, divided into two groups, positive axioms (as for the implicational fragment of **R**, see Dunn and Restall [21, Section 1.5], where alternative equivalent axiomatizations are discussed), and negative axioms.*

positive axioms:
  self implication $\vdash_{\mathcal{H}_{rc}} \varphi \to \varphi$
  prefixing $\vdash_{\mathcal{H}_{rc}} (\varphi \to \psi) \to [(\chi \to \varphi) \to (\chi \to \psi)]$
  contraction $\vdash_{\mathcal{H}_{rc}} [\varphi \to (\varphi \to \psi)] \to (\varphi \to \psi)$
  permutation $\vdash_{\mathcal{H}_{rc}} [\varphi \to (\psi \to \chi)] \to [\psi \to (\varphi \to \chi)]$

negative axioms: *First, note that because of the absence of conjunction from the object-language, biconditional '$\leftrightarrow$' cannot be directly defined. Below, I use axioms schemes with the notation $\varphi \leftrightarrow \psi$ as an abbreviation of pairs of axions schemes $\varphi \to \psi$ and $\psi \to \varphi$. Compare those negative axioms with the standard Relevance logic negative axioms mentioned in Example 5.2.25.*

  double negation $\vdash_{\mathcal{H}_{rc}} \neg\neg\varphi \leftrightarrow \varphi$
  negating implication $\vdash_{\mathcal{H}_{rc}} \neg(\varphi \to \psi) \leftrightarrow (\varphi \to \neg\psi)$

with the single inference rule

$$\frac{\varphi \quad \varphi \to \psi}{\psi} \; (MP)$$

Suppose, first, that derivations were defined as usual for Hilbert-like systems[2], namely, sequences of formulas where each member is an assumption, or an axiom, or the result of an application of (MP) to two

---
[2] Referred to as 'protoproofs' in Avron [4].

5.3 Axiomatic definition of $\mathcal{L}_{rc}$

earlier members in the list. Denote by $\vdash_{\mathcal{H}_{rc}} \Gamma : \varphi$ the derivability in $\mathcal{H}_{rc}$ of φ from a multi-set of open assumptions Γ. Here Γ, φ and Γ_1, Γ_2 mean multi-set union. Again, when Γ is empty, φ is a thesis (formal theorem) of $\mathcal{L}_{rc}$.

Clearly, this notion of derivation cannot be shown equivalent to $\mathcal{N}_{rc}$-derivations, as, in contrast to $\mathcal{N}_{rc}$-derivations, it ignores the relevance of assumptions to the derived conclusion. See the discussion in Dunn and Restall [21, Section 2.1].

To make the two notions of derivations compatible for comparison, I redefine the notion of $\mathcal{H}_{rc}$-derivation by incorporating into it the tracing of used assumptions, as suggested in Dunn and Restall [21]. Thus, each assumption is flagged with a *fresh* flag, say i, and if an assumption occurs more than once in a derivation, it always occurs with the same index. Thus, assumptions can be still considered as forming a set. Furthermore, (MP) is modified to

$$\frac{\varphi_\alpha \quad \varphi \rightarrow \psi_\beta}{\psi_{\alpha \cup \beta}} \ (MP^r)$$

If, again by convention, we suppose that the assumptions Γ are flagged by $\{1 \cdots n\}$ (for some $n \geq 0$), then the conclusion has to be flagged by $\hat{n}$, having used all the assumptions, as in $\mathcal{N}_{rc}$. Denote this relevant deducibility notion by $\vdash^r_{\mathcal{H}_{rc}} \Gamma : \varphi_{\hat{n}}$.

5.3.2 Deductive equivalence of $\mathcal{H}_{rc}$ and $\mathcal{N}_{rc}$

Theorem 5.3.3. *For every Γ and φ:*

$$\vdash^r_{\mathcal{H}_{rc}} \Gamma : \varphi_{\hat{n}} \text{ iff } \vdash_{\mathcal{N}_{rc}} \Gamma : \varphi_{\hat{n}} \qquad (5.3.26)$$

Proof. Without loss of generality, assume the open assumptions Γ are flagged for use with the same indices in both derivations. In $\mathcal{H}_{rc}$, axioms can be considered as flagged with $\emptyset$.

- To show that $\vdash^r_{\mathcal{H}_{rc}} \Gamma : \varphi_{\hat{n}}$ implies $\vdash_{\mathcal{N}_{rc}} \Gamma : \varphi_{\hat{n}}$, I show that all the $\mathcal{H}_{rc}$ axioms are derivable (from no open assumptions) in $\mathcal{N}_{rc}$. The derivations for the positive axioms are rather standard and omitted (see, for example, Example 5.2.2.4). I show the derivations for the negative axioms.

  double negation:

  $$\dfrac{\dfrac{\dfrac{[\varphi]_1}{\varphi_1}(Ass)}{\neg\neg\varphi_1}(dni)}{\varphi\rightarrow\neg\neg\varphi}(\rightarrow^+I^1) \qquad \dfrac{\dfrac{\dfrac{[\neg\neg\varphi]_1}{\neg\neg\varphi_1}(Ass)}{\varphi_1}(dne)}{\neg\neg\varphi\rightarrow\varphi}(\rightarrow^+I^1) \qquad (5.3.27)$$

  negating implication: The derivations are those of Boethius' theses in (5.2.10).

- To show that $\vdash_{\mathcal{N}_{rc}} \Gamma : \varphi_{\hat{n}}$ implies $\vdash^r_{\mathcal{H}_{rc}} \Gamma : \varphi_{\hat{n}}$, assume $\vdash_{\mathcal{N}_{rc}} \Gamma : \varphi_{\hat{n}}$. The proof is by induction on the last rule applied in the $\mathcal{N}_{rc}$-derivation. Again, only the negative rules are of interest.

  ($\rightarrow^- I$): In this case, φ is $\neg(\psi\rightarrow\chi)$ for some ψ, χ. The premise
  $$[\psi]_i$$
  $$\vdots$$
  of this application of ($\rightarrow^- I$) is $(\neg\chi)_{\hat{n}\cup i}$. By the induction hypothesis on the premise, $\vdash^r_{\mathcal{H}_{cr}} \Gamma, \psi : (\neg\chi)_{\hat{n}\cup i}$. By the deduction-theorem, $\vdash^r_{\mathcal{H}_{cr}} \Gamma : (\psi\rightarrow\neg\chi)_{\hat{n}}$, and by the negating application axiom and (MP^r), $\vdash^r_{\mathcal{H}_{cr}} \Gamma : (\neg(\psi\rightarrow\chi))_{\hat{n}}$.

  ($\rightarrow^- E$): In this case, φ is $\neg\psi$ for some ψ, and the premises of the rule are $\neg(\chi\rightarrow\psi)_\alpha$ and χ_β, for some χ. By the induction hypothesis on the premises, $(*)$ $\vdash^r_{\mathcal{H}_{rc}} \Gamma_1 : \neg(\chi\rightarrow\psi)_{\hat{n}_1}$ and $(**)$ $\vdash^r_{\mathcal{H}_{rc}} \Gamma_2 : \chi_{\hat{n}_2}$, where $\Gamma_1\Gamma_2 = \Gamma$ and $\hat{n}_1\cup\hat{n}_2 = \hat{n}$. From $(*)$ and the negating implication axiom, we get by (MP^r) $(***)$ $\vdash^r_{\mathcal{H}_{rc}} \Gamma : (\chi\rightarrow\sim\psi)_{\hat{n}_2}$. Finally, from $(**)$ and $(***)$ we get by (MP^r) $\vdash^r_{\mathcal{H}_{rc}} \Gamma :\sim \psi_{\hat{n}}$.

  (dni), (dne) Obvious and omitted.

  $\square$

Corollary 5.3.9 (the relevant-connexive deduction-theorem).

(DT_{rc}^+) $i \notin \hat{n}$ and $\vdash_{\mathcal{H}_{rc}}^r \Gamma, \varphi_i : \psi_{\hat{n} \cup i}$ iff $\vdash_{\mathcal{H}_{rc}}^r \Gamma : (\varphi \to \psi)_{\hat{n}}$ \hfill (5.3.28)

(DT_{rc}^-) $i \notin \hat{n}$ and $\vdash_{\mathcal{H}_{rc}}^r \Gamma, \varphi_i : \neg\psi_{\hat{n} \cup i}$ iff $\vdash_{\mathcal{H}_{rc}}^r \Gamma : \neg(\varphi \to \psi)_{\hat{n}}$ \hfill (5.3.29)

Proof. First, for (DT_{rc}^+), assume (1) $\vdash_{\mathcal{H}_{rc}}^r \Gamma, \varphi_i : \psi_{\hat{n} \cup i}$ for some $i \notin \hat{n}$. By Theorem 5.3.3, (1) holds iff (2) $\vdash_{\mathcal{N}_{rc}} \Gamma, \varphi_i : \psi_{\hat{n} \cup i}$. By Theorem 5.2.2, (2) holds iff (3) $\vdash_{\mathcal{N}_{rc}} \Gamma : (\varphi \to \psi)_{\hat{n}}$. So, again by Theorem 5.3.3, (3) holds iff (4) $\vdash_{\mathcal{H}_{rc}}^r \Gamma : (\varphi \to \psi)_{\hat{n}}$, which establishes the result.

The proof of (DT_{rc}^-) is similar and omitted. □

5.4 Conclusion

In this chapter, a connexive extension $\mathcal{L}_{rc}$ of the Relevance logic $\mathbf{R}_\to$ was presented. It is defined by means of a natural-deduction system, and a deductively equivalent axiomatic system is presented too. The goal of such an extension is to produce a logic with stronger connection between the antecedent and the consequent of an implication.

An interesting question, deserving further research, is about the possibility of devising a connexive extension for a fuller Relevance logic, including conjunction and disjunction. Following Avron [5], I believe that the *extensional* (boolean) conjunction and disjunction (known also as multiplicative connectives) do not fit Relevance logic, and their *intensional* counterparts, fusion and fission (known also as additive connectives) are preferred. See Avron [5] for an extensive discussion; see also Mares [69].

However, once the relevant conditional and negation are replaced by a their connexive counterparts, the usual definitions of the intensional connectives cannot be applied anymore. For example, in $\mathbf{R}_{\to,\neg}$, one can define fusion by

$$\varphi \circ \psi =^{df.} \neg(\varphi \to \neg\psi)$$

This renders fusion identical to implication!

$$\neg(\varphi \to \neg\psi) \text{ iff } (\varphi \to \neg\neg\psi) \text{ iff } \varphi \to \psi$$

So, some more suitable definition of fusion is called for.

Chapter 6

An application of Connexive logics

6.1 Introduction

In this chapter, I present an application of Connexive logics. That is, uses of a Connexive logic for matters other than a mere investigation of their intrinsic properties or relationship to other logics: defining a *connexive class theory* instead of the standard class theory defined by using first-order Classical logic. This definition first appeared in Francez [35]. The main tool used for this application is the notion of *restricted quantification* (known also as binary quantification), where the classical material conditional on which such quantification is based is replaced by a connexive conditional.

Two other attempted applications were shown less successful.

- An attempt to define *arithmetic* (even a very weak one) over Connexive logics was shown unsuccessful by Ferguson [30].

- An attempt to define (a version of) *set theory* over Connexive logics was shown unsuccessful by Wiredu [127].

6.2 Connexive class theory

6.2.1 Restricted quantification

I want to investigate the meaning of *restricted quantification* (RQ), when expressed in a first-order *Connexive logic*. The general forms of RQ are the universal generalization of the conditional and the existential generalization of the conjunction, respectively:

$$(RQ_\forall): \forall x.\varphi \to \psi \qquad (RQ_\exists): \exists x.\varphi \wedge \psi \qquad (6.2.1)$$

Here '$\forall$', '$\exists$', '$\to$' and '$\wedge$' are some *generic* quantifiers (universal and existential), and connectives (conditional, conjunction), taken from some first-order logic, and φ, ψ typically have a free occurrence of x.

I focus here on $RQ_\forall$. The existential $RC_\exists$ is usually taken as the dual of $RC_\forall$, and its interpretation follows from that of $RC_\forall$ by dualising. Thus, there is a dependence of the meaning of $RQ_\forall$ on the meaning of the conditional it embeds. For a similar study of RQ induced by *Relevance logics* see Beall, Brady, Hazen, Priest and Restall [7].

The best known application of $RQ_\forall$ in Classical logic is its definitional role in a logical scheme of defining classes and operations thereon, delineated in Section 6.3. So, I will adopt this tool as the guiding line for the study of $RQ_\forall$.

Another axis along which this study advances is a critical inspection of a natural correspondence between class-theoretic notions and classical

6.2 Connexive class theory

logical notions, as depicted below. The classes here are classes of models (or "cases"), if logical consequence is viewed more generally as in *logical pluralism*,

logic	class theory
contradiction	empty class
tautology	universal class
entailment	class inclusion
equivalence	class equality

The question is, how much can this correspondence be adopted to paraconsistent/paracomplete logics?

The following picture emerges from this correspondence.

A first-order logic's paraconsistency and paracompleteness can be judged by the class-theory it induces.

It might be expected that in a connexive class theory:

- The partial account of paraconsistency and paracompleteness corresponds to the class-theoretic properties that the empty class is not a subclass of every class and not every class is a subclass of the universal class.

- The null account of paraconsistency and paracompleteness corresponds to the class-theoretic properties that the empty class is a subclass of *no* class (except itself), and no class is a subclass of the universal class (except itself).

A major characteristic of connexivity of a first-order logic corresponds to a class-theoretic property: a class is not **a subclass of its complement**. This characteristic leads to the adoption of the partial accounts in those case that a connexive restricted quantification does indeed lead to a connexive class theory.

I first delineate the class-definition scheme in first-order Classical logic in Section 6.3. Then, in Section 6.5 the investigation of connexive $RC_\forall$ with Wansing's 1st-order Connexive logic QC (a first-order extension of the propositional Connexive logic C), the only first-order connexive logic proposed so far, concluding that its induced class-definition scheme does not meet the yardstick for a connexive class-theory. In Section 3.3 I continue the investigation with QP_S, a 1st-order extension of Priest's Connexive logic P_S, concluding that it also does not meet the yardstick for a connexive class-theory, because implication and consequence are based on the null account for paraconsistency and paracompleteness.

Then, in Section 6.7 I introduce a variant QP_S^p of Priest's QP_S, that is based on the partial account for paraconsistency/paracompleteness, and show that it *does* meet the yardstick for a connexive class-theory. The Section ends with some conclusions.

6.3 Defining classes in first-order Classical logic

As a preparatory step, in this section I recapitulate the way classes and operations thereon are defined in Classical logic. Note that the issue is the expressivity of class-related definitions in the logic, and *not* (naive or axiomatic) *set theory*, ZFC or other, taking 'ϵ' as a primitive operation. Throughout the chapter, it is assumed that 'ϵ' is *not* a primitive relational symbol in the first-order logics considered.

All the definitions are relative to a classical model $\mathcal{M} = \langle D, I \rangle$, where, as usual, D is some non-empty set, the domain of the model, and I is an

6.3 Defining classes in first-order Classical logic

interpretation function, interpreting constants and predicate symbols. I assume that each element of D, say d, serves as a name for itself. I use $\varphi[x := d]$ as the result of substituting d for every free occurrence of x in φ, occasionally denoted also as $\varphi(d)$. The usual classical definition of satisfaction in a model, consequence and validity are assumed.

class formation: Every *open* formula, say $\varphi(x)$, having a single free variable, say x, defines a class $\widehat{\varphi(x)}_{\mathcal{M}}$, the *extension* of $\varphi(x)$ in $\mathcal{M}$, by

$$\widehat{\varphi(x)}_{\mathcal{M}} =^{df.} \{d \in D \mid \mathcal{M} \vDash \varphi[x := d]\} \quad (6.3.2)$$

This way of defining a class is essentially[1] *class abstraction*.

One can extend the class formation way also to *closed* formulas. Closed formulas satisfy that $\varphi[x := d]$ is identical to φ, and $\hat{\varphi}_{\mathcal{M}}$ can be taken to define the universal class D in case $\mathcal{M} \vDash \varphi$, or define the empty class $\varnothing$ in case $\mathcal{M} \nvDash \varphi$.

class membership: For $d \in D$, $d \in \widehat{\varphi(x)}_{\mathcal{M}}$ iff $\mathcal{M} \vDash \varphi[x := d]$.

Boolean operations: The operations of intersection, union and complementation are represented by, respectively, conjunction, disjunction and negation of open formulas. Thus,

$$\widehat{\varphi(x)}_{\mathcal{M}} \cap \widehat{\psi(x)}_{\mathcal{M}} =^{df.} \widehat{\varphi(x) \wedge \psi(x)}_{\mathcal{M}}$$

$$\widehat{\varphi(x)}_{\mathcal{M}} \cup \widehat{\psi(x)}_{\mathcal{M}} =^{df.} \widehat{\varphi(x) \vee \psi(x)}_{\mathcal{M}} \quad (6.3.3)$$

$$\overline{\widehat{\varphi(x)}_{\mathcal{M}}} =^{df.} \widehat{\neg \varphi(x)}_{\mathcal{M}}$$

boundary classes: These are obtained by quantification.

- If $\mathcal{M} \vDash \forall x. \neg \varphi(x)$, that is $\varphi(x)$ is identically false in $\mathcal{M}$, then $\varnothing =^{df.} \widehat{\varphi(x)}_{\mathcal{M}}$, the *empty* class.

---

[1] Recall that the object-language is assumed *not* to contain '$\in$' as a primitive symbol, so it is impossible to form Russell's paradoxical set.

- If $\mathcal{M} \models \forall x.\varphi(x)$, that is $\varphi(x)$ is identically true in $\mathcal{M}$, then $\overline{\varphi(x)}_\mathcal{M} = D$, the *universal* class.

Note that boundary classes are uniquely defined by open formulas up to logical equivalence. That is, if both $\overline{\varphi(x)}_\mathcal{M} = \emptyset$ and $\overline{\psi(x)}_\mathcal{M} = \emptyset$, then $\mathcal{M} \models \forall x.\varphi(x) \equiv \psi(x)$. Similarly for universal class definitions.

class inclusion: It is here that restricted quantification has its effect.

$$\overline{\varphi(x)}_\mathcal{M} \subseteq \overline{\psi(x)}_\mathcal{M} \text{ iff } \mathcal{M} \models \forall x.\varphi(x) \supset \psi(x) \qquad (6.3.4)$$

Since the generic conditional is taken in Classical logic as the truth-functional material implication '$\supset$', restricted quantification $RQ_\forall$ can be in understood as expressing class inclusion.

From (6.3.4) we get the standard definition of class equality, expressing the *extensionality* of classes defined by this scheme..

$$\overline{\varphi(x)}_\mathcal{M} = \overline{\psi(x)}_\mathcal{M} \text{ iff } \mathcal{M} \models \forall x.\varphi(x) \equiv \psi(x) \qquad (6.3.5)$$

Corollary 6.3.9. *Some notable consequences of this definition of classical classes, to be contrasted with what follows, are:*

1. $\overline{\varphi(x)}_\mathcal{M} \cap \overline{\neg\varphi(x)}_\mathcal{M} = \emptyset, \quad \overline{\varphi(x)}_\mathcal{M} \cup \overline{\neg\varphi(x)}_\mathcal{M} = D$

2. $\emptyset \subseteq \overline{\varphi(x)}_\mathcal{M} \subseteq D$

3. $\emptyset \subseteq \overline{\emptyset} \, [= D]$

4. $\overline{\varphi(x)}_\mathcal{M} \subseteq \overline{\psi(x)}_\mathcal{M}$ *implies* $\overline{\neg\psi(x)}_\mathcal{M} \subseteq \overline{\neg\varphi(x)}_\mathcal{M}$

6.4 Connexivity and class inclusion

6.4.1 Connexive algebra

The impact of taking the conditional as connexive (and not as material) on the expression of class definition, in particular class inclusion and boundary classes, was first brought to the front by McCall [72]. The latter paper starts with a dialogue between a student and a teacher, at the beginning of which the student expresses the following intuition about class inclusion:

> Student: no class can be included in its complement.

The teacher then reminds the student about the empty class ∅, that by the Cantor's paradise definition of inclusion is an exception to that intuition. The student then complains about the definition of inclusion, and McCall developed an algebraic theory CA (connexive algebra), parallel to Boolean algebras, in which the student's intuition indeed obtains.

The theory CA is formulated in a kind of hybrid of a propositional logic and an algebra. It has variables $a, b, \cdots$ ranging over elements of a carrier, and two kinds of connectives:

logical (proposition forming) : '⊃' (conditional) and '∼' (negation), forming compound propositions, with the usual definition of conjunction '·', disjunction '∨' and biconditional '≡' in terms of the conditional and the negation.

algebraic (term forming) : class intersection '$\alpha \cap \beta$' (with $\alpha \cap \beta$ written as $\alpha\beta$) and class complementation α'. The union is defined as usual by $(\alpha'\beta')'$. There are two defined constants: 0 (denoting the empty class) and 1 (denoting the universal class), defined, respectively, as $0 =^{df.} aa'$ and $1 =^{df.} 0'$.

$$(x \subset y) \wedge (y \subset z) \supset (x \subset z) \qquad (6.4.6)$$

$$x \subset y \supset xz \subset yz \qquad (6.4.7)$$

$$x(y \cup z) \subset ay \cup az \qquad (6.4.8)$$

$$x(yz) \subset (xy)z \qquad (6.4.9)$$

$$(xy' \subset 0) \wedge (x \not\subset 0) \wedge (y' \not\subset 0) \supset x \subset y \qquad (6.4.10)$$

$$(y' \subset x') \supset (x \subset y) \qquad (6.4.11)$$

$$xx' \subset yy' \qquad (6.4.12)$$

$$x \subset xx \qquad (6.4.13)$$

$$x \subset x'' \qquad (6.4.14)$$

$$(x \subset 0) \supset (0 \subset x) \qquad (6.4.15)$$

$$(0 \subset x) \supset (x \subset 0) \qquad (6.4.16)$$

$$x \not\subset x' \qquad (6.4.17)$$

Figure 6.1: Axioms of CA

Atomic propositions have the form $\alpha \subset \beta$, denoting class inclusion. Equality is defined as $\alpha = \beta =^{df.} (\alpha \subset \beta) \cdot (\beta \subset \alpha)$.

There are two inference rules: Modus Ponens (MP) and substitution of terms for variables.

The theory has non-logical axioms, presented in Figure 6.1. Those axioms are like the axioms of a Boolean algebra, but with the additional non-Boolean ones: (6.4.16) and (6.4.17). The axiom in (6.4.17) is a clear algebraic analog to the connexivity characteristic A_1 (in (2.2.1)), and is the axiom expressing the students's intuition in the initial dialog. The axioms certainly do not conform to the relationships in Corollary

6.4 Connexivity and class inclusion

6.3.9. The axiom in (6.4.16) expresses the upshot of McCall's CA, that in statements about classes, the empty class and the universal class have to be treated separately. In particular:

- Classes that are neither empty nor universal have the same properties as the classically defined classes.

- The empty class *is not* a subclass of every class – only of itself.

  This is the class-theoretic facet of paraconsistency, according to the natural correspondence specified above, induced by connexivity. It is a partial paraconsistent entailment by Priest's classification.

- The universal class *does not* contain every class as a subclass – only itself.

  This is the class-theoretic facet of paracompleteness, according to the natural correspondence specified above, induced by connexivity. It is a partial paracomplete entailment by Priest's classification.

It is plausible to believe that a more natural setting for the development of a connexive class-theory is at a first-order level of a Connexive logic. Two such logics, Wansing's QC, and a first-order variant of Priest's P_S (in two guises), are delineated in the coming sections, and the question to be explored is the following: can a an application of a class-definition scheme expressed in each of those logics, in analogy to that used by Classical logic, produce a connexive class theory in the spirit of McCall's connexive algebra CA?

An answer to this question sheds light on the meaning of $RC_\forall$ in Connexive logics.

6.5 QC - Wansing's first-order connexive logic

I next consider Wansing's QC [119] as a candidate for constituting a first-order Connexive logic underlying the current goal.

6.5.0.1 Axiomatic definition of QC

The object-language of C is extended with a denumerable set of *terms* over variables and constants, ranged over by t, as well as predicate symbols. Let T denote the collection of all terms, and CT - the collection of all closed terms. Atomic and general formulas are defined as usual. The quantifiers are taken as mutually dual. The axiomatic presentation of C in Section 3.2.1 is extended with the following axioms for quantification. All substitutions are assumed clash-free, and $\varphi[x := t]$ is abbreviated to $\varphi(t)$.

$$\vdash_{QC} \varphi(t) \to \exists x.\varphi(x)$$
$$\vdash_{QC} \forall x.\varphi(x) \to \varphi(t) \tag{6.5.18}$$

In addition, the following two quantification rules (my names) are included.

$$\frac{\varphi \to \psi(x)}{\varphi \to \forall x.\psi(x)} (\forall) \qquad \frac{\varphi(x) \to \psi}{\exists x.\varphi(x) \to \psi} (\exists) \tag{6.5.19}$$

where in ($\forall$) x is not free in φ and in ($\exists$) x is not free in ψ.

By using any valid formula as φ in ($\forall$), we get the useful derived rule for universal generalization:

$$\frac{\psi(x)}{\forall x.\psi(x)} (\forall I) \tag{6.5.20}$$

6.5 QC - Wansing's first-order connexive logic

The first-order connexive characteristics in (2.3.15) and (2.3.16) are immediately derived by $(\forall I)$ from their propositional counterparts in (2.2.1) and (2.2.3).

For example, $\vdash_{QC} \neg((\forall x.\varphi(x)) \to \neg \forall x.\varphi(x))$ is a closed instance of A_1.

Also, from the open instance

$$\vdash_{QC} \neg(\varphi(x) \to \neg \varphi(x))$$

of A_1, we get by $(\forall I)$

$$\vdash_{QC} \forall x. \neg(\varphi(x) \to \neg \varphi(x))$$

6.5.0.2 Complete models for QC

The main idea here is that the verification conditions for the universal quantifier and the falsification conditions for the existential quantifier are also dynamic.

A QC-model $\mathcal{M}$ is a tuple $\mathcal{M} = \langle W, \leq, \Delta, \mathcal{D}, v^+, v^- \rangle$, adding two components[2] to C-models:

- Δ is a set of terms s.t. $CT \subseteq \Delta \subseteq T$, intended for use in the clauses for verification and falsification for open formulas.

- $\mathcal{D}$ maps points to subsets of Δ s. t.:

    1. For every $t \in W$, $CT \subseteq \mathcal{D}_t$.
    2. If $s \leq t$ then $\mathcal{D}_s \subseteq \mathcal{D}_t$.

---
[2]Note the font difference between $\mathcal{D}$, the assignment of a domain to each point, and D, the domain in a classical model and the universal set in the class-definition scheme.

- Both v^+ and v^- respect $\mathcal{D}$:
  If $t \in v^{+/-}(P(a_1, \cdots, a_n))$, then $\{a_1, \cdots, a_n\} \subseteq \mathcal{D}_t$.

The truth/falsity support relations of C are extended[3] as follows.

$\mathcal{M}, t \models^+ p(x)$ iff $t \in v^+(p(x))$ [implying $x \in \mathcal{D}_t$]

$\mathcal{M}, t \models^- p(x)$ iff $t \in v^-(p(x))$ [idem]

$\mathcal{M}, t \models^+ \forall x.\varphi$ iff   for every $s \geq t$: for every $d \in \mathcal{D}_s$ $\mathcal{M}, s \models^+ \varphi(d)$

$\mathcal{M}, t \models^- \forall x.\varphi$ iff   for some $d \in \mathcal{D}_s$ $\mathcal{M}, t \models^- \varphi(d)$

$\mathcal{M}, t \models^+ \exists x.\varphi$ iff   for some $d \in \mathcal{D}_s$ $\mathcal{M}, t \models^+ \varphi(d)$

$\mathcal{M}, t \models^- \exists x.\varphi$ iff   for every $s \geq t$: for every $d \in \mathcal{D}_s$ $\mathcal{M}, s \models^- \varphi(d)$

(6.5.21)

Like C, QC is both paraconsistent and paracomplete, as the valuations $v^{+/-}$ are required to be neither disjoint nor exhaustive.

Validity and logical consequence are defined analogously to Definition 3.2.1.

Let us return to the basic question, what is expressed by $RQ_\forall$ in QC? The ingredient in the model-theory to take account of again is the additional relativisation to points introduced by the universal quantifier, on top of the relativisation induced by the embedded connexive conditional.

---

[3]The clauses for truth/falsity support for *open formulas* are not listed explicitly in Wansing [119], but can be deduced from the accompanying description. I thank Heinrich Wansing for clarifications regarding this point.

6.5 QC - Wansing's first-order connexive logic

We have the following:

$\mathcal{M}, t \models^+ \forall x.\varphi(x) \to \psi(x)$ iff $(\forall s \geq t)(\forall d \in \mathcal{D}_s): \mathcal{M}, s \models^+ \varphi(d) \to \psi(d)$

$$\text{iff} \quad \begin{array}{l}(\forall s \geq t)(\forall d \in \mathcal{D}_s)(\forall r \geq s): \\ \text{if } \mathcal{M}, r \models^+ \varphi(d) \text{ then } \mathcal{M}, r \models^+ \psi(d)\end{array}$$

$$\text{iff} \quad \begin{array}{l}(\forall s \geq t)(\forall d \in \mathcal{D}_s): \\ \text{if } \mathcal{M}, s \models^+ \varphi(d) \text{ then } \mathcal{M}, s \models^+ \psi(d)\end{array} \quad (6.5.22)$$

The extra quantification induced by '$\forall r$' has no effect.

Similarly,

$\mathcal{M}, t \models^- \forall x.\varphi(x) \to \psi(x)$ iff $(\exists d \in \mathcal{D}_t): \mathcal{M}, t \models^- \varphi(d) \to \psi(d)$

$$\text{iff} \quad \begin{array}{l}(\exists d \in \mathcal{D}_t)(\forall s \geq t): \\ \text{if } \mathcal{M}, s \models^+ \varphi(d) \text{ then } \mathcal{M}, s \models^- \psi(d)\end{array} \quad (6.5.23)$$

6.5.1 Defining classes in QC

In this section, I investigate how the class-definition scheme is reflected in QC, and how much is this reflection analogous to that in Classical logic as in Section 6.3. In particular, can we obtain classes defined in QC as models of McCall's CA?

By the structure of a QC model, there are two major observations here:

1. Definitions are relative to a point $t \in W$, resulting in an intensional notion of a class.

2. An open formula $\varphi(x)$ naturally defines in a model $\mathcal{M}$ at each

point t *two* extensions: a positive one $\widehat{\varphi(x)}^+_{\mathcal{M},t}$ and a negative one one, $\widehat{\varphi(x)}^-_{\mathcal{M},t}$, the latter to be called $\varphi(x)$'s *coextension*.

This is due to the separate definition of $\models^+$ and $\models^-$ and the relativization of both to points $t \in W$.

class formation:

$$\widehat{\varphi(x)}^+_{\mathcal{M},t} =^{df.} \{d \in CT \mid \mathcal{M}, t\models^+ \varphi[x := d]\} \tag{6.5.24}$$

$$\widehat{\varphi(x)}^-_{\mathcal{M},t} =^{df.} \{d \in CT \mid \mathcal{M}, t\models^- \varphi[x := d]\} \tag{6.5.25}$$

Note that $\Delta - CT \subseteq Var$, a collection of variables possibly in $\mathcal{D}_t$, does not take part in the definition of the above extensions. Only closed terms (i.e., CT) do.

Example 6.5.31 (a running example). *Suppose the object-language contains a unary predicate P. Consider a model $\mathcal{M}_\mathcal{N}$ in which $W_\mathcal{N} = \mathcal{N}$, (the points are the natural numbers, naturally ordered), $CT = \{0, 1, 2, 3\}$ and $\mathcal{D}_{\mathcal{N},n} = CT \cup \{x\} = \{0, 1, 2, 3\} \cup \{x\}$, for every point $n \geq 0$, where $x \in \Delta_\mathcal{N}$. Let*

$$\begin{cases} v^+_\mathcal{N}(P(n)) = \mathcal{N} & \text{for } n \in \{0, 1\} \\ v^-_\mathcal{N}(P(n)) = \mathcal{N} & \text{for } n \in \{1, 2\} \end{cases}$$

This example is uniform over all points. That is, every point n supports the truth of $P(0)$ and $P(1)$ and supports the falsity of $P(1)$ and $P(2)$. That is:

- *$P(1)$ is supported both for truth and for falsity at every point.*
- *$P(3)$ is neither supported for truth nor for falsity at any point.*

6.5 QC - Wansing's first-order connexive logic

We have

$$\widehat{P(x)}^+_{M_N,n} = \{0,1\} \quad \widehat{P(x)}^-_{M_N,n} = \{1,2\} \quad \text{for every point } n$$

A natural question arising at this point is, what is the relationship between $\varphi(x)$'s extension $\widehat{\varphi(x)}^+_{M,t}$ and its coextension $\widehat{\varphi(x)}^-_{M,t}$, for any $t \in W$?

On the one hand, in view of the interpretation of '$\vDash^-$' as support of falsity, one is naturally led to the identification of the coextension $\widehat{\varphi(x)}^-_{M,t}$ with the *complement* (at t) of the extension $\widehat{\varphi(x)}^+_{M,t}$. This idea is reinforced by the clauses for negation, by which

$$\mathcal{M}, t \vDash^- \varphi(x) \text{ iff } \mathcal{M}, t \vDash^+ \sim \varphi(x) \tag{6.5.26}$$

That is, $\widehat{\varphi(x)}^-_{M,t} = \sim \widehat{\varphi(x)}^+_{M,t}$. So, let us turn this into the definition of the complement, keeping the analogy with the classical definition of the complement.

$$\overline{\widehat{\varphi(x)}^+_{M,t}} =^{df.} \widehat{\varphi(x)}^-_{M,t} \ [= \sim \widehat{\varphi(x)}^+_{M,t}] \tag{6.5.27}$$

In addition, this view leads to an *involutive* complementation.

On the other hand, in view of non-disjointness and non-exhaustiveness of support of truth and support of falsity, we may have models $\mathcal{M}$ (for example, $\mathcal{M}_N$ in example 6.5.31 above, see below) and points t therein in which $\widehat{\varphi(x)}^+_{M,t}$ and $\widehat{\varphi(x)}^-_{M,t}$ are themselves neither disjoint nor exhaustive, in contrast both to classically defined complementation and to complementation in CA.

Example 6.5.32 (continued). *Note that in our example:*

- $\widehat{P(x)}^+_{M_N,n} \cap \widehat{P(x)}^-_{M_N,n} = \{1\}$: *non-disjointness (at every point).*

- $\widehat{P(x)}^+_{M_N,n} \cup \widehat{P(x)}^-_{M_N,n} = \{0,1,2\} \subset \{0,1,2,3\}$: *non-exhaustiveness (at every point)*

Clearly, such a definition of a complement is not like that of a Boolean complement.

class membership: Continuing the analogy to classical class definition, whereby membership is determined by predication, we again relativize to points. However, both positive predication and negative predication, defined separately, need to be accounted for. This leads naturally to consider also *anti-membership*, denoted[4] by '$\notin$'.

For $d \in CT$:

$$d \in_t \overline{\varphi(x)}^+_{\mathcal{M},t} \text{ iff } \mathcal{M}, t \models^+ \varphi[x := d] \quad (6.5.28)$$

$$d \notin_t \overline{\varphi(x)}^+_{\mathcal{M},t} \text{ iff } \mathcal{M}, t \models^- \varphi[x := d] \quad (6.5.29)$$

An element d can satisfy *both* $d \in_t \overline{\varphi(x)}^+_{\mathcal{M},t}$ and $d \in_t \overline{\varphi(x)}^-_{\mathcal{M},t}$. Similarly, d can satisfy both $d \notin_t \overline{\varphi(x)}^+_{\mathcal{M},t}$ and $d \notin_t \overline{\varphi(x)}^-_{\mathcal{M},t}$. This is another facet of the non-disjointness and non-exhaustiveness of a class and its complement, already noted above, not in accordance with the axioms of CA.

Example 6.5.33 (continued). *In our example, at every point n:*

$$1 \in_n \overline{P(x)}^+_{\mathcal{M}_\mathcal{N},n} \text{ and } 1 \in_n \overline{P(x)}^-_{\mathcal{M}_\mathcal{N},n} [= \overline{\overline{P(x)}^+_{\mathcal{M}_\mathcal{N},n}}] \quad (6.5.30)$$

$$3 \notin_n \overline{P(x)}^+_{\mathcal{M}_\mathcal{N},n} \text{ and } 3 \notin_n \overline{P(x)}^-_{\mathcal{M}_\mathcal{N},n} [= \overline{\overline{P(x)}^+_{\mathcal{M}_\mathcal{N},n}}] \quad (6.5.31)$$

boundary classes: What about empty and universal classes?

Suppose again we keep the analogy with he classical definitions of the boundary classes, taking into account the additional dynamization introduced by the universal quantification.

---

[4]Not to be confused with the use of this symbol as denoting the complement of the membership relation.

6.5 QC - Wansing's first-order connexive logic

Definition 6.5.10 (empty class).

$$\widetilde{\varphi(x)}^+_{M,t} = \varnothing \quad \text{iff } M, t \models^+ \forall x. \sim \varphi(x)$$

$$\text{iff } (\forall s \geq t)(\forall d \in \mathcal{D}_s) M, s \models^+ \sim \varphi(d)$$

or, equivalently $\quad$ iff $(\forall s \geq t)(\forall d \in \mathcal{D}_s) M, s \models^- \varphi(d)$
$$\tag{6.5.32}$$

Definition 6.5.11 (universal class).

$$\widetilde{\varphi(x)}^+_{M,t} = D \,[= CT] \quad \text{iff } M, t \models^+ \forall x. \varphi(x)$$

$$\text{iff } (\forall s \geq t)(\forall d \in \mathcal{D}_s) M, s \models^+ \varphi(d) \tag{6.5.33}$$

The problem, again, is that those definitions do not express "real emptiness" and "real universality", because of the possibility that for some $d \in \mathcal{D}_s$, either both of $M, s \models^+ \varphi(d)$ and $M, s \models^- \varphi(d)$ are satisfied, or neither is.

In an "extreme" variation of the model $\mathcal{M}_\mathcal{N}$ (from Example 6.5.31), in which at every point $t \in W$, for every $d \in CT$, both $v^+(P(d)) = \mathcal{N}$ and $v^-(P(d)) = \mathcal{N}$, that is $P(d)$ is supported both for truth and for falsity at every point, we get that $\varnothing = D$!

Boolean operations: Those are also pointwise, and are defined both for positive and for negative extensions.

$$\widetilde{\varphi(x) \wedge \psi(x)}^+_{M,t} =^{df.} \widetilde{\varphi(x)}^+_{M,t} \cap \widetilde{\psi(x)}^+_{M,t}$$

$$\widetilde{\varphi(x) \wedge \psi(x)}^-_{M,t} =^{df.} \widetilde{\varphi(x)}^-_{M,t} \cup \widetilde{\psi(x)}^-_{M,t}$$
$$\tag{6.5.34}$$

$$\widetilde{\varphi(x) \vee \psi(x)}^+_{M,t} =^{df.} \widetilde{\varphi(x)}^+_{M,t} \cup \widetilde{\psi(x)}^+_{M,t}$$

$$\widetilde{\varphi(x) \vee \psi(x)}^-_{M,t} =^{df.} \widetilde{\varphi(x)}^-_{M,t} \cap \widetilde{\psi(x)}^-_{M,t}$$

class inclusion: Let us turn (6.5.22) into the definition of connexive class inclusion.

Definition 6.5.12 (connexive class inclusion).

$$\overline{\varphi(x)}^+_{M,t} \subseteq \overline{\psi(x)}^+_{M,t} \text{ iff } (\forall s \geq t)(\forall d \in \mathcal{D}_s) \text{ if } \mathcal{M}, s \models^+ \varphi(d) \text{ then } \mathcal{M}, s \models^+ \psi(d) \quad (6.5.35)$$

Consequently, we get for class equality the following definition.[5]

$$\overline{\varphi(x)}^+_{M,t} = \overline{\psi(x)}^+_{M,t} \text{ iff } (\forall s \geq t)(\forall d \in \mathcal{D}_s) \ \mathcal{M}, s \models^+ \varphi(d) \text{ iff } \mathcal{M}, s \models^+ \psi(d) \quad (6.5.36)$$

What happens in the special case in which $\psi(x)$ is taken as $\sim \varphi(x)$, a central case for connexive class-theory?

From the characteristic QA_1, we have that

$$\overline{\varphi(x)}^+_{M,t} \not\subseteq \overline{\overline{\varphi(x)}}^+_{M,t} \quad (6.5.37)$$

This *is* in accordance to the CA Axiom in (6.4.17).

In conclusion, QC does not lead naturally to CA models by the connexive class-definition scheme, not so much because of the connexive reading of $RC_\forall$ (which yields the connexively favourable (6.5.37)), as for its strong negation. QC departs from Classical logic more strongly then what seems to be required to affect only boundary classes. The problem of the impossibility to express "real emptiness" and "real universality" seems to be related to the problem of the inexpressibility of 'just true' and 'just false' in logics with truth-value *gluts* and *gaps*. See, for example, Beall [6].

---

[5]One might consider strengthening the definition of class equality by demanding, in addition, the equality of the coextensions of φ and ψ at every $s \geq t$. This will not change much if CA is the goal.

6.6 Priest's first-order connexive logics

I now turn to the first-order extension of propositional Connexive logic introduced in by Priest [96]. As we will see, two natural variants of this extension emerge, differing in their suitability for defining a connexive class theory.

6.6.1 First-order QP_S

In Priest [96], the definition of a first-order extension of P_S is stated as obvious and not given. I present one here, to be referred to as QP_S. Ferguson [30] also considers a first-order extension of Priest's P_N with a null account of logical consequence in the context of connexive arithmetic.

First, like in QC, the object-language of P_S is extended with denumerable sets of *terms*, over variables and constants, ranged over by t, as well as predicate symbols. Let T denote the collection of all terms, and CT - the collection of all closed terms. Atomic and general formulas are again defined as usual.

Models for QP_S contain a domain, a non-empty set D, invariant across worlds. There are two kinds of assignment functions: a world independent variable-assignment σ, assigning each variable an element in the domain, intended for interpreting open formulas, and a collection of extension-assignment functions v_w assigning to $R(x_1, \cdots, x_n)$ its extension in w, $v_w[\![R(x_1, \cdots, x_n)]\!] \subseteq D^n$. Let $\sigma[x := d]$ be the variable-assignment function assigning d to x and coinciding with σ on any other variable.

The following clauses are added to the specification of the satisfaction relation, that is now relative both to a world and to a global variable-assignment function.

$$\mathcal{M}, w, \sigma \vDash R(x_1, \cdots, x_n) \quad \text{iff} \quad \langle \sigma[\![x_1]\!], \cdots, \sigma[\![x_n]\!] \rangle \in v_w[\![R(x_1, \cdots, x_n)]\!]$$

$$\mathcal{M}, w, \sigma \vDash \forall x.\varphi(x) \quad \text{iff} \quad \begin{array}{l}(\forall d \in D)(\forall w' \in W): \\ \mathcal{M}, w', \sigma[x = d] \vDash \varphi(x)\end{array}$$

(6.6.38)

Thus, the dynamization of the universal quantifier is employed here too. This is crucial for obtaining a CA by the class-definition scheme.

Satisfaction is naturally extended to:

$$\mathcal{M}, w \vDash \varphi \text{ iff } \mathcal{M}, w, \sigma \vDash \varphi \text{ for every } \sigma \qquad (6.6.39)$$

Validity is still defined as in (3.3.12), namely satisfaction at g.

Let us see the how $RC_\forall$ looks like in QP_S. Consider a model $\mathcal{M}$.

$$\mathcal{M} \vDash \forall x.\varphi(x) \to \psi(x) \quad \text{iff} \quad \mathcal{M}, g \vDash \forall x.\varphi(x) \to \psi(x)$$

$$\text{iff} \quad (\forall d \in D)(\forall w' \in W) \mathcal{M}, w' \vDash \varphi(d) \to \psi(d)$$

$$\text{iff} \quad \begin{array}{l}(\forall d \in D)(\forall w' \in W) \\ \exists w'' \in W: \mathcal{M}, w'' \vDash \varphi(d) \text{ and} \\ \exists w'' \in W: \mathcal{M}, w'' \vDash \neg \psi(d) \text{ and} \\ \forall w'' \in W: \begin{array}{l}\text{if } \mathcal{M}, w'' \vDash \varphi(d) \\ \text{then } \mathcal{M}, w'' \vDash \psi(d)\end{array}\end{array}$$

(6.6.40)

It is not hard to see that the characteristics of first-order connexivity, specified in Section 2, are valid in QP_S. For example, to see that QA_1 is valid, consider any model $\mathcal{M}$, and assume that for some world w in it and some $d \in D$, it holds that $\mathcal{M}, w \vDash \varphi(d) \to \neg\varphi(d)$. Then, just apply Priest's argument at the end of Section 3.3.

The other characteristics are established similarly.

Example 6.6.34. *Consider*

$$\forall x.\varphi(x) \to \varphi(x)$$

6.6 Priest's first-order connexive logics

For this formula to be valid in a model $\mathcal{M}$ of QP_S, $\varphi(x)$ should neither hold for *every* $d \in \mathcal{D}$, nor for *no* $d \in \mathcal{D}$.

6.6.2 Defining classes in QP_S

In this section, I investigate how the class-definition scheme is applied in QP_s, and, again, how much is this application analogous to that in Classical logic as in Section 6.3. In particular, can we obtain classes defined in QP_S by this scheme as models of McCall's CA?

By the structure of QP_S-models, there is again a natural *single* extension associated with any open formula $\varphi(x)$. Furthermore, the distinguished world g can serve as an anchor, allowing to weaken the class definition dependence on worlds.

As the application of the class definition scheme below shows, we are back to a classical-like definition, but in the distinguished world g.

class formation:

$$\widehat{\varphi(x)}_{\mathcal{M}} =^{df.} \{d \in CT \mid \mathcal{M}, g \models \varphi[x := d]\} \quad (6.6.41)$$

class membership: For $d \in D$, $d \in \widehat{\varphi(x)}_{\mathcal{M}}$ iff $\mathcal{M}, g \models \varphi[x := d]$.

Boolean operations:

$$\widehat{\varphi(x) \wedge \psi(x)}_{\mathcal{M}} =^{df.} \widehat{\varphi(x)}_{\mathcal{M}} \cap \widehat{\psi(x)}_{\mathcal{M}}$$

$$\widehat{\varphi(x) \vee \psi(x)}_{\mathcal{M}} =^{df.} \widehat{\varphi(x)}_{\mathcal{M}} \cup \widehat{\psi(x)}_{\mathcal{M}} \quad (6.6.42)$$

$$\widehat{\neg \varphi(x)}_{\mathcal{M}} =^{df.} \overline{\widehat{\varphi(x)}_{\mathcal{M}}}$$

boundary classes: Here the anchoring with satisfaction at g has its effect.

- If $M\models \forall x.\neg\varphi(x)$, that is, for every $d \in D$, $\varphi(d)$ is false in M at g, then $\widehat{\varphi(x)}_M = \emptyset$.
  Note that by this definition

$$\widehat{\varphi(x)}_M \cap \widehat{\neg\varphi(x)}_M = \emptyset \qquad (6.6.43)$$

and

$$\widehat{\varphi(x)}_M \cup \widehat{\neg\varphi(x)}_M = D \qquad (6.6.44)$$

- If $M\models \forall x.\varphi(x)$, that is, for every $d \in D$, $\varphi(d)$ is true in M at g, then $\widehat{\varphi(x)}_M = D$.

class inclusion: It is here that the connexive restricted quantification has its effect.

$$\widehat{\varphi(x)}_M \subseteq \widehat{\psi(x)}_M \text{ iff } M\models \forall x.\varphi(x)\to\psi(x) \qquad (6.6.45)$$

That is, iff $M, g \models \forall x.\varphi(x)\to\psi(x)]$.

Let $\mathcal{C}_M$ denote the collection of classes defined over M. That is, the collection of all $\widehat{\varphi(x)}_M$, when $\varphi(x)$ varies over all open formulas (with only x free) in the object-language.

The next proposition establishes the unsuitability of QP_S's connexive class theory as a suitable definitional tool to match McCall's connexive algebra.

Proposition 6.6.31 (first-order connexive class theory). *For no QP_S-model M is $\mathcal{C}_M$ a CA (a connexive algebra).*

Proof: While $a \not\subseteq a'$ holds immediately by QA_1, there is a problem due to the reliance of the definition of inclusion on the null account. From the axioms, we have that $(*)$ $0 \subset 0$ is a theorem of CA. Thus, in the class theory induced by QP_S, $\emptyset \subset \emptyset$ should hold. Recall that in every model M, $\widehat{\varphi(x)\wedge\neg\varphi(x)}_M = \emptyset$. Thus, for $(*)$ to hold, both

$$M\models \forall x.\varphi(x)\wedge\neg\varphi(x)\to\varphi(x)\wedge\neg\varphi(x)$$

and

$$\mathcal{M} \models \forall x.\varphi(x) \wedge \neg\varphi(x) \rightarrow \psi(x) \wedge \neg\psi(x)$$

have to hold. This entails (by the first clause of the definition of satisfaction of implication – the null account!) that for all $d \in D$ and $w' \in W$, there exists a $w'' \in W$ such that:

$$\mathcal{M}, w'' \models \varphi(d) \wedge \neg\varphi(d)$$

which is impossible.

6.7 Another variant of Priest's connexive logic

Since the culprit of the failure of QP_S's class theory to produce a CA is the definition of the satisfaction of the implication, requiring the existence of a model for the antecedent (by the null account), changing this definition may lead to the required result.

I now define a variant P_S^p of P_S, based on the *partial* account. The partiality I assume renders the implication, and the respective logical consequence relation, *reflexive* (in a broad sense): *The only conclusions of a contradiction are all other contradictions, including itself, and a tautology is a conclusion only of other tautologies, including itself.*

To that effect, the clause in (3.3.11) pertaining to implication is weakened to

$$\mathcal{M}, w \vDash \varphi \rightarrow \psi \quad \text{iff}$$
either
(1.) $[\forall w' \in W : \mathcal{M}, w' \not\vDash \varphi$ and
$\forall w' \in W : \mathcal{M}, w' \not\vDash \psi]$
or
(2.) $[\forall w' \in W : \mathcal{M}, w' \vDash \varphi$ and (6.7.46)
$\forall w' \in W : \mathcal{M}, w' \vDash \psi]$
or
(3.) $[\exists w' \in W : \mathcal{M}, w' \vDash \varphi$ and
$\exists w' \in W : \mathcal{M}, w' \vDash \neg \psi$ and
$\forall w' \in W :$ if $\mathcal{M}, w' \vDash \varphi$ then $\mathcal{M}, w' \vDash \psi]$

Thus, reflexivity (in this broad sense) is *imposed* on the otherwise null account!

As for logical consequence, a similar weakening is applied to (3.3.13), obtaining a *partial* account by imposing reflexivity.

$$\Gamma \vDash_{P_S^p} \varphi \quad \text{iff} \quad \begin{array}{l} \text{either } \varphi \in \Gamma \\ \text{or} \\ [\exists \mathcal{M} : \mathcal{M} \vDash \Gamma \text{ and} \\ \exists \mathcal{M} : \mathcal{M} \vDash \neg \varphi \text{ and} \\ \forall \mathcal{M} : \text{if } \mathcal{M} \vDash \Gamma \text{ then } \mathcal{M} \vDash \varphi] \end{array} \quad (6.7.47)$$

Indeed, those are partial account paraconsistency and partial account paracompleteness!

The first-order extension QP_S^p is defined analogously to the definition of QP_S.

It is not hard to see that the connexive first-order characteristics remain intact by this modification. However, it does affect the induced class theory. We now finally have:

Theorem 6.7.4 (first-order connexive class theory). *For every QP_S^p-model $\mathcal{M}$, $\mathcal{C}_\mathcal{M}$ is a CA (a connexive algebra).*

6.7 Another variant of Priest's connexive logic

Proof of Theorem 6.7.4 Consider an arbitrary model QP_S^p-model $\mathcal{M}$ and $w \in W$. I have to show that all of the connexive axioms in figure 6.1 are satisfied by $\mathcal{C}_\mathcal{M}$. I will show only some, making clear how the reasoning proceeds. Suppose $x = \widetilde{\varphi(z)}$ and $y = \widetilde{\psi(z)}$

$xx' \subset yy'$: This requires $\mathcal{M}, w \models \forall z.(\varphi(z) \land \neg\varphi(z)) \to (\psi(z) \land \neg\psi(z))$.
Since both the antecedent and consequent are unsatisfiable, the result follows from clause (1.) of (6.7.46).

This settles the problem with the obstacle we have encountered in QP_S, the invalidity of $\emptyset \subset \emptyset$. It now holds due to the imposed reflexivity.

The dual property about the universal set, namely $(1 \subset 1)$, follows by contraposition.

$x \subset xx$: I distinguish between three cases.

$x = 0$: In this case, the requirement is
$$\mathcal{M}, w \models \forall z.((\varphi(z) \land \neg\varphi(z)) \to ((\varphi(z) \land \neg\varphi(z)) \land (\varphi(z) \land \neg\varphi(z))))$$
Again, both the antecedent and consequent are unsatisfiable, and the result follows from clause (1.) of (6.7.46).

$x = 1$: In this case, the requirement is
$$\mathcal{M}, w \models \forall z.((\varphi(z) \lor \neg\varphi(z)) \to (\varphi(z) \lor \neg\varphi(z)) \land (\varphi(z) \lor \neg\varphi(z)))$$
Since both the antecedent and consequent are valid, the result follows from clause (2.) of (6.7.46).

$x \neq 0 \land x \neq 1$: In this case, both $\varphi(x)$ and $\psi(x)$ are contingent, and the result follows from clause (3.) of (6.7.46).

$x \subset x''$: Again, the same three cases need to be distinguished.

$x = 0$: In this case, the requirement is
$$\mathcal{M}, w \models \forall z.((\varphi(z) \land \neg\varphi(z)) \to \neg\neg(\varphi(z) \land \neg\varphi(z)))$$
Both the antecedent and consequent are unsatisfiable, and the result follows from clause (1.) of (6.7.46).

x = 1: In this case, the requirement is

$$\mathcal{M}, w \models \forall z.((\varphi(z) \vee \neg \varphi(z)) \to \neg \neg (\varphi(z) \vee \neg \varphi(z)))$$

Both the antecedent and consequent are valid, and the result follows from clause (2.) of (6.7.46).

$x \neq 0 \wedge x \neq 1$: In this case, both $\varphi(x)$ and $\psi(x)$ are contingent, and the result follows from clause (3.) of (6.7.46).

$x \notin x'$: In this case, the requirement is

$$\mathcal{M}, w \models \forall z. \neg(\varphi(z) \to \neg \varphi(z))$$

That is,

$$\mathcal{M}, w \not\models \forall z. \varphi(z) \to \neg \varphi(z)$$

Clearly, if $\varphi(z)$ is unsatisfiable, then $\neg \varphi(z)$ is a tautology, and if $\varphi(z)$ is a tautology, then $\neg \varphi(z)$ is unsatisfiable. So, the first two clauses of (6.7.46) cannot apply. Also, the third conjunct of the third clause of (6.7.46) prevents that clause to apply either, so the required non-satisfaction of the quantified conditional follows.

6.8 Conclusions

In this section, I have studied the meaning of universal restricted quantification $RC_\forall$ in three first-order Connexive logics: Wansing's QC and two first-order variants of Priest's P_S. The main tool employed in this study is the role of the connexive $RC_\forall$ when taking part in a first-order class-definition scheme. As the model of connexive class-theory I adopted McCall's connexive algebra (CA), presented algebraically at a propositional level.

The conclusions reached are:

6.8 Conclusions

- QC does not lead naturally to CA by the class-definition scheme. The structure of its models involves ingredients that interfere with this scheme, mainly the appeal to strong negation.

- QP_S, based on the null account of implication and logical consequence, does not lead either to CA by the class-definition scheme.

- However, a variant QP_S^p, based on a partial account (imposing reflexivity of implication and of logical consequence), does lead to CA by that scheme.

- The important ingredients in the structure of models for first-order Connexive logics, leading to the validity of the first-order connexivity characteristics, are the dynamization of both the connexive conditional and the connexive universal quantifier, the two components of the connexive restricted quantification $RC_\forall$.

A possible response to the above observations can be that CA is not the right definition of a connexive class-theory. An alternative class-theory might be lacking any boundary classes altogether, with membership being more loose. Alternatively, it may have boundary classes, but those are "isolated", not taking part in the inclusion relation at all, not even including themselves.[6] I am not aware at this stage of any uses of a connexive class-theory that may decide this issue. The advantage I do see, though, is the restoration of the null account of paraconsistency/paracompleteness, that I consider preferable for independent reasons.

An interesting continuation of this line of research is an investigation of a different connexive $RQ_\forall$, which embeds a connexive *dual implication* (or *coimplication*) instead of embedding a connexive conditional. Such a connective is introduced in Wansing's propositional bi-

---

[6]In such an algebra, equality has to be primitive, not defined via mutual inclusion as in CA.

connexive logic [120], still awaiting its extension to a first-order variant.

Chapter 7

Connexivity and modality

7.1 Introduction

Modal logics, with their modal operators '□' (necessity) and '◊' (possibility) have several ways of being associated with connexivity. In this chapter, I review the main ways such associations were suggested in the literature.

One way to relate modality with connexivity, which we encountered already, is to use the modal operators as a way of weakening the connexive axioms to obtain humble connexivity. This is discussed in Section 7.2.

Another approach to relating modality and connexivity is a proposal by Estrada-González [25], where the modal operators are *identified* with certain connexive theses. For example,

$$\Diamond \varphi =^{df.} \neg(\varphi \to \neg\varphi)$$

Necessity is the obtained by duality. This approach is presented in Section 7.3.

Yet another, very different and more common, approach is discussed in some detail in Wansing [122], who devotes a whole section (4.6) to connexive modal logics. As is evident from the opening sentences in that section, the main interest is in *modal extension of connexive logics*, and various embeddings induced by such extension. Thus, connexivity takes effect in producing instances of the connexive axioms with modal formulas. For example,

$$\vdash \neg(\Box\varphi \to \neg\Box\varphi) \tag{7.1.1}$$

$$\vdash \neg(\neg\Box\varphi \to \Box\varphi) \tag{7.1.2}$$

From the point of view of the Bochum plan, this modal extensions of connexive logics on the modified falsification condition of the conditional, already inducing connexivity of the underlying logic.
There is no additional modified falsification condition of the modal operators themselves.
This approach is presented in Section 7.4.

I propose to employ the Bochum plan in a stronger way: modify also the falsification conditions of the modal operators '$\Box$' and '$\Diamond$', obtaining formal theorems that *are not* instances of the underlying connexive axioms. This is presented in Section 7.5.

7.2 Weakening the connexive axioms

As stated above, a way the modal operators can be related to connexivity is in connection with humble connexivity (cf. Section 4.2). If the modal operators are in the object-language, other ways of implementing Kapsner's idea of restricting the scope of connexivity by weakening the connexive axioms to possible antecedent are

$$\begin{aligned} A_1^{\Diamond,\to} &: \ \vdash \Diamond\varphi \to \neg(\varphi \to \neg\varphi) \\ A_2^{\Diamond,\to} &: \ \vdash \Diamond\varphi \to \neg(\neg\varphi \to \varphi) \end{aligned} \tag{7.2.3}$$

or Kapsner's preferred formulation

$$\begin{aligned}A_1^{\Diamond,\vdash} &: \Diamond\varphi\vdash\neg(\varphi\to\neg\varphi)\\ A_2^{\Diamond,\vdash} &: \Diamond\varphi\vdash\neg(\neg\varphi\to\varphi)\end{aligned} \quad (7.2.4)$$

Similar modal weakening applies also to the B_i connexive axioms.

A similar weakening is endorsed by Iacona [48], who claims that Lewis' strict conditional, defined as $\Box(\varphi\supset\psi)$, validates this weakened version of the connexive axioms, and hence is a more adequate conditional than the connexive one.

In addition to excluding impossible antecedents, Iacona excludes also from the scope of connexivity conditionals with a necessary consequent, basically for similar reasons.

Another use of the modal operators in relation to connexivity is weakening the way the conditional is negated. In [22], Egré and Politzer investigate empirically the negation scheme of the conditional

$$\neg(\varphi\to\psi)\equiv\varphi\to\Diamond\neg\psi$$

Such a scheme is motivated by probabilistic considerations, into which I will not enter here.

7.3 Defining the modal operators

Estrada-González in [25] gives a definition of the modal operators of possibility '$\Diamond$' and necessity '$\Box$' using the connexive connectives. His definitions follow ideas of Lewis and Langford [61], relating possibility to *self-consistency*: φ is possible iff it is self-consistent.

Lewis and Langford take consistency between φ and ψ, denoted by $\varphi\circ\psi$, to be defined as

$$\varphi\circ\psi =^{df.} \neg(\varphi\to\neg\psi)$$

which leads to

$$\Diamond\varphi =^{df.} \varphi \circ \varphi = \neg(\varphi \rightarrow \neg\varphi) \qquad (7.3.5)$$

And by assuming the traditional duality between possibility and necessity, the latter is defines as

$$\Box\varphi = \neg\Diamond\neg\varphi = \neg\varphi \rightarrow \varphi \qquad (7.3.6)$$

Estrada-González [25] then investigates the resulting modal logics, in particular with relation to *possibilism* the view that *everything is possible (and nothing is necessary)*. Notably, the resulting modalities can be defined by three-valued truth-tables.

This approach is further studied in Nicolás-Francisco [80].

7.4 Connexive modal logics

In this section, I consider ways of obtaining a *Connexive modal logic*, the object-language of which is extended to include the modal operators. Such logics should validate instances of the connexive axioms in which the formulas feature also modal operators, such as the following instance of A_1:

$$\vdash \neg(\Box\varphi \rightarrow \neg\Box\varphi)$$

There is a distinction between (at least) two approaches for obtaining such logics.

- *Extend* an underlying non-modal Connexive logic by adding the modal operators to the object-language and extending the logic's definition so as to allow instances of the connexive axioms involving formulas with modalities.

- Extend an underlying non-connexive Modal logic.

7.4.1 Wansing's CK connexive modal logic

One Connexive modal logic is Wansing's CK from [119], a Connexive modal logic which is a connexive analog to the smallest Modal logic K satisfying $\Box(\varphi\to\psi)\to(\Box\varphi\to\Box\psi)$. The logic CK is obtained as an extension of Wansing's Connexive logic C (see Section 3.2). The extension involves augmenting the frames (and models based on them) with a binary accessibility relation R (related in a certain way to '$\leq$'), and adding the following clauses for support of truth and of falsity for the modal operators.

$$\mathcal{M}, t \vDash^+ \Box\varphi \text{ iff } \forall u \geq t \forall v. uRv \text{ implies } \mathcal{M}, v \vDash^+ \varphi$$
$$\mathcal{M}, t \vDash^- \Box\varphi \text{ iff } \exists u. tRu \text{ and } \mathcal{M}, u \vDash^- \varphi$$

$$\mathcal{M}, t \vDash^+ \Diamond\varphi \text{ iff } \exists u. tRu \text{ and } \mathcal{M}, u \vDash^+ \varphi$$
$$\mathcal{M}, t \vDash^- \Diamond\varphi \text{ iff } \forall u \geq t \forall v. uRv \text{ implies } \mathcal{M}, v \vDash^- \varphi$$

The semantics is complete w.r.t. an axiomatization presented in [119].

7.4.1.1 Connexive modal extensions of FDE

In [81], Odintsov, Skurt and Wansing introduce several modal extensions of the four-valued logic FDE. I will delineate one of them. I first briefly review FDE itself, then a (non-connexive) modal extension of it $KFDE$ (from Priest [98]) and then BK^-, a connexive extension of $KFDE$.

7.4.1.2 Review of FDE

FDE, the logic of *first-degree entailment* (i.e., without an embeddable conditional) is a four-valued logic over the connectives $\{\neg, \wedge, \vee\}$ and the truth-values $\mathcal{V} = \{t, b, n, f\}$, interpreted as follows:

- t and f are the traditional truth-values representing *true* and *false*, respectively.

- b represents *both true and false*, a *glut*.

- n represents *neither true and false*, a *gap*.

Equivalently, $\mathcal{V}$ can be viewed as the isomorphic power set of $\{1,0\}$, namely $\{\{1\},\{1,0\},\{0\},\varnothing\}$, and both views will be used as convenient.

FDE was first introduced by Belnap [8, 9] and further studied by Dunn [20]. A more extensive overview can be found in Omori and Wansing [86].

The object-language of FDE constitutes of $\{\neg, \wedge, \vee\}$.

The logic is usually presented model-theoretically, using a two-valued relational[1] model-theory due to Dunn [20], justifying the above interpretation of the four truth-values.

Definition 7.4.13 (*FDE* interpretations). *An FDE interpretation is a relation $r \subseteq \mathbf{At} \times \{0,1\}$, relating atomic formulas to (classical) truth-value. The interpretation is extended to compound formulas as follows.*

$$\neg\varphi\, r\, 1 \text{ iff } \varphi\, r\, 0$$
$$\neg\varphi\, r\, 0 \text{ iff } \varphi\, r\, 1$$
$$\varphi \wedge \psi\, r\, 1 \text{ iff } \varphi\, r\, 1 \text{ and } \psi\, r\, 1$$
$$\varphi \wedge \psi\, r\, 0 \text{ iff } \varphi\, r\, 0 \text{ or } \psi\, r\, 0$$
$$\varphi \vee \psi\, r\, 1 \text{ iff } \varphi\, r\, 1 \text{ or } \psi\, r\, 1$$
$$\varphi \vee \psi\, r\, 0 \text{ iff } \varphi\, r\, 0 \text{ and } \psi\, r\, 0$$

□

Remark 7.4.10. *1. Formulas are r-related to 0 or to 1 independently. In particular, $\varphi\, r\, 1$ may be different from $\varphi \not{r}\, 0$, i.e., being related to 1 is not the same as not being related to 0.*

---
[1] There are also four-valued truth-tables for FDE

7.4 Connexive modal logics

2. *r can relate a formula φ both to 0 and to 1, or to neither. Hence the interpretation of $\mathcal{V}$.*

□

Definition 7.4.14 (*FDE* logical consequence). *ψ is an FDE logical consequence of Γ, denoted by $\Gamma \vDash_{FDE} \psi$, iff for every interpretation r: if $\varphi r 1$ for every $\varphi \in \Gamma$, then $\psi r 1$.* □

Clearly, *FDE* is a paraconsistent logic Since both $\varphi\ r\ 1$ and $\neg\varphi\ r\ 1$ may hold without $\psi\ r\ 1$ holding, so in such a case $\varphi, \neg\varphi \nvDash_{FDE} \psi$.

7.4.1.3 A non-connexive modal extension of *FDE*

Next, the modal extension *KFDE* of *FDE*, due to Priest [98], is delineated. The object-language of *KFDE* is extended with the modal operators '$\Box$' and '$\Diamond$'.

The logic is presented model-theoretically.

Definition 7.4.15 (*KFDE* models). *A KFDE model is a tuple $\langle W, R, v^+, v^- \rangle$, where:*

- *W is a non-empty set of evaluation points (called also (information) states), ranged over by w, u.*

- *$R \subseteq W \times W$ is a (binary) accessibility relation on W.*

- *$v^+, v^- : \mathcal{P} \Rightarrow 2^W$ are functions mapping formulas to sets of points.*

Next, the support of truth/falsity *of a formula by a point, denoted $w \Vdash^+$*

φ and $w \Vdash^- \varphi$, respectively, are defined recursively.

$$w \Vdash^+ p \text{ iff } w \in v^+[\![p]\!], \text{ for } p \in \mathcal{P}$$
$$w \Vdash^- p \text{ iff } w \in v^-[\![p]\!], \text{ for } p \in \mathcal{P}$$
$$w \Vdash^+ \neg\varphi \text{ iff } w \Vdash^- \varphi$$
$$w \Vdash^- \neg\varphi \text{ iff } w \Vdash^+ \varphi$$
$$w \Vdash^+ \varphi \wedge \psi \text{ iff } w \Vdash^+ \varphi \text{ and } w \Vdash^+ \psi$$
$$w \Vdash^- \varphi \wedge \psi \text{ iff } w \Vdash^- \varphi \text{ or } w \Vdash^- \psi$$
$$w \Vdash^+ \varphi \vee \psi \text{ iff } w \Vdash^+ \varphi \text{ or } w \Vdash^+ \psi$$
$$w \Vdash^- \varphi \vee \psi \text{ iff } w \Vdash^- \varphi \text{ and } w \Vdash^- \psi$$
$$w \Vdash^+ \Box\varphi \text{ iff } \forall u: R(w,u) \text{ implies } u \Vdash^+ \varphi$$
$$w \Vdash^- \Box\varphi \text{ iff } \exists u: R(w,u) \text{ and } u \Vdash^- \varphi$$
$$w \Vdash^+ \Diamond\varphi \text{ iff } \exists u: R(w,u) \text{ and } u \Vdash^+ \varphi$$
$$w \Vdash^- \Diamond\varphi \text{ iff } \forall u: R(w,u) \text{ implies } u \Vdash^- \varphi$$

$\square$

Clearly, support of truth and support of falsity are relativisations of the FDE r-relatedness to points.

Definition 7.4.16 ($KFDE$ logical consequence). *ψ is an $KFDE$ logical consequence of Γ, denoted by $\Gamma \vDash_{KFDE} \psi$, iff for every model $\mathcal{M}$ and every $w \in W$: if $w \Vdash^+ \varphi$ for every $\varphi \in \Gamma$, then $w \Vdash^+ \psi$.* $\square$

7.4.1.4 A connexive modal extension of FDE

I now delineate one of the several connexive extensions of $KFDE$ presented in Odintsov, Skurt and Wansing [81], a Connexive modal logic called cBK^-. The object-language is extended with a conditional '$\rightarrow_c$' with the following conditions for support of truth and falsity.

$$w \Vdash^+ \varphi \rightarrow_c \psi \text{ iff } w \Vdash^+ \varphi \text{ implies } w \Vdash^+ \psi$$
$$w \Vdash^- \varphi \rightarrow_c \psi \text{ iff } w \Vdash^+ \varphi \text{ implies } w \Vdash^- \psi$$

We see again the familiar falsification condition of the connexive conditional, familiar from thee Bochum plan.

The proofs of the connexive axioms in cBK^- are similar to the proofs of those axioms in C and omitted.

Proposition 7.4.32 (negation inconsistency of cBK^-).

Proof. The following instance of the negation of A_2 is provable in cBK^-

$$\vdash_{cBK^-} \neg(\varphi \to_c \varphi) \to_c (\varphi \to_c \varphi)$$

We have that $w \Vdash^+ \neg(\varphi \to_c \varphi) \to_c (\varphi \to_c \varphi)$ iff:
If $w \Vdash^+ \varphi$ implies $\Vdash^- \varphi$, then $w \Vdash^+ \varphi$ implies $\Vdash^+ \varphi$, which clearly holds.

$\square$

However, obviously cBK^- is not trivial, so it is paraconsistent.

Odintsov, Skurt and Wansing [81] show that cBK^- is sound and complete w.r.t. a tableau proof-system they present, as well as some further result transcending the scope of the current presentation.

7.5 Francez' Connexive modal logic $CS5$

7.5.1 Introduction

In this section, I present the Connexive modal logic $CS5$, with modified falsification conditions both for the conditional *and* for the modalities. I consider here only the implication-negation-modalities fragment. The logic is again motivated by certain pattern of intonational stress, as described in Section 7.6.5.

I first consider only the necessity operator '$\Box$', deferring the possibility operator '$\Diamond$' to Section 7.6.6.

The modified falsification condition for '$\Box$' satisfies the axiom[2]

$$\neg\Box\varphi \equiv \Box\neg\varphi \qquad (7.5.7)$$

In Section 7.6.5, I present some natural language uses of negating necessity as support of (7.5.7), rendering the latter non-arbitrary.

The modal logic that I take as a point of departure is $S5$, avoiding considerations of accessibility relations in the model-theory. However, my definitional tools is a *hyper-sequent calculus SCS5*, a modification of the hyper-sequent calculus for $S5$ as presented by Poggiolesi [95], with modified rules for negating both '$\Box$' and '$\Diamond$'.

In particular, I am interested in obtaining, under the modified falsification condition of '$\Box$', the following:

$$\vdash_{SCS5} \neg(\Box\varphi \rightarrow \Box\neg\varphi) \qquad \vdash_{SCS5} \neg(\Box\neg\varphi \rightarrow \Box\varphi) \qquad (7.5.8)$$

which are not instances of the A_i axioms, neither are they theorems of $S5$.

7.6 The connexive modal logic $CS5$

Notation: A *sequent* is of the form $\Gamma : \Delta$, and a hyper-sequent is of the form $\Gamma_1 : \Delta_1 \mid \cdots \mid \Gamma_n : \Delta_n$ for some $n \geq 1$. Hyper-sequents are ranged over by $\mathcal{G}$. Note that ':' is the empty sequent.

---

[2]This axiom is known from non-connexive modal logics with a functional accessibility relation in its Kripke frames.

7.6.1 The sequent calculus $SCS5$

The rules of $SCS5$ are presented[3] in Figure 7.1. The structural rules of Weakening and Contraction are assumed both at the level of sequents and at the level of hyper-sequents. Also, Cut is assumed for sequents. As I do not use them in any example derivations below, and I am not dealing with their admissibility, I do not present them explicitly.

7.6.2 Derivability of the $S5$ axioms

Clearly, because of the modified falsification condition of the conditional, not all classical propositional validities are derivable. However, the modal axioms of $S5$ remain derivable, as shown below.

K: $\vdash_{SCS5} \Box(\varphi \to \psi) \to (\Box\varphi \to \Box\psi)$:

$$\cfrac{\cfrac{\cfrac{\cfrac{\cfrac{\cfrac{\varphi : \varphi \quad \psi : \psi}{\varphi, \varphi \to \psi : \psi} (\to L)}{\Box(\varphi \to \psi) : \mid \varphi : \psi} (\Box L_2)}{\Box(\varphi \to \psi), \Box\varphi : \mid : \psi} (\Box L_2)}{\Box(\varphi \to \psi), \Box\varphi : \Box\psi} (\Box R)}{\Box(\varphi \to \psi) : (\Box\varphi \to \Box\psi)} (\to R)}{: \Box(\varphi \to \psi) \to (\Box\varphi \to \Box\psi)} (\to R)$$

T: $\vdash_{SCS5} \Box\varphi \to \varphi$:

$$\cfrac{\cfrac{\varphi : \varphi}{\Box\varphi : \varphi} (\Box L_1)}{: \Box\varphi \to \varphi} (\to R)$$

---

[3]In Poggiolesi [95], the modalized formula is repeated in the premise, to obtain invertibility of the rules. Since this issue is orthogonal to connexivity, I present the simpler version of the rules without this repetition

Initial sequents:
$$\mathcal{G} \mid \varphi, \Gamma : \Delta, \varphi \qquad (7.6.9)$$

Propositional rules:
$$\frac{\mathcal{G} \mid \Gamma : \Delta, \varphi \quad \mathcal{G} \mid \psi, \Gamma : \Delta}{\mathcal{G} \mid \varphi \to \psi, \Gamma : \Delta} \; (\to L) \qquad (7.6.10)$$

$$\frac{\mathcal{G} \mid \varphi, \Gamma : \Delta, \psi}{\mathcal{G} \mid \Gamma : \Delta, \varphi \to \psi} \; (\to R) \qquad (7.6.11)$$

$$\frac{\mathcal{G} \mid \Gamma : \Delta, \varphi \quad \mathcal{G} \mid \neg\psi, \Gamma : \Delta}{\mathcal{G} \mid \neg(\varphi \to \psi), \Gamma : \Delta} \; (\neg\to L) \qquad (7.6.12)$$

$$\frac{\mathcal{G} \mid \varphi, \Gamma : \Delta, \neg\psi}{\mathcal{G} \mid \Gamma : \Delta, \neg(\varphi \to \psi)} \; (\neg\to R) \qquad (7.6.13)$$

$$\frac{\mathcal{G} \mid \varphi, \Gamma : \Delta}{\mathcal{G} \mid \neg\neg\varphi, \Gamma : \Delta} \; (\neg\neg L) \quad \frac{\mathcal{G} \mid \Gamma : \Delta, \varphi}{\mathcal{G} \mid \Gamma : \Delta, \neg\neg\varphi} \; (\neg\neg R) \qquad (7.6.14)$$

Modal rules (for '$\Box$'):
$$\frac{\mathcal{G} \mid \varphi, \Gamma : \Delta}{\mathcal{G} \mid \Box\varphi, \Gamma : \Delta} \; (\Box L_1) \quad \frac{\mathcal{G} \mid \Gamma : \Delta \mid \varphi, \Sigma : \Theta}{\mathcal{G} \mid \Box\varphi, \Gamma : \Delta \mid \Sigma : \Theta} \; (\Box L_2) \qquad (7.6.15)$$

$$\frac{\mathcal{G} \mid \Gamma : \Delta \mid : \varphi}{\mathcal{G} \mid \Gamma : \Delta, \Box\varphi} \; (\Box R) \qquad (7.6.16)$$

$$\frac{\mathcal{G} \mid \neg\varphi, \Gamma : \Delta}{\mathcal{G} \mid \neg\Box\varphi, \Gamma : \Delta} \; (\neg\Box L_1) \quad \frac{\mathcal{G} \mid \Gamma : \Delta \mid \neg\varphi, \Sigma : \Theta}{\mathcal{G} \mid \neg\Box\varphi, \Gamma : \Delta \mid \Sigma : \Theta} \; (\neg\Box L_2) \qquad (7.6.17)$$

$$\frac{\mathcal{G} \mid \Gamma : \Delta \mid : \neg\varphi}{\mathcal{G} \mid \Gamma : \Delta, \neg\Box\varphi} \; (\neg\Box R) \qquad (7.6.18)$$

Figure 7.1: The sequent calculus $SCS5$ for $CS5$

7.6 The connexive modal logic $CS5$

B: $\vdash_{SCS5} \Box\varphi \to \Box\Box\varphi$:

$$\cfrac{\cfrac{\cfrac{\cfrac{: \mid \varphi : \varphi}{\Box\varphi : \mid : \mid : \varphi} \, (\Box L_2)}{\Box\varphi : \mid : \Box\varphi} \, (\Box R)}{\Box\varphi : \Box\Box\varphi} \, (\Box R)}{: \Box\varphi \to \Box\Box\varphi} \, (\to R)$$

5: $\vdash_{SCS5} \neg\Box\neg\varphi \to \Box\neg\Box\neg\varphi$

$$\cfrac{\cfrac{\cfrac{\cfrac{: \mid \neg\neg\varphi : \neg\neg\varphi}{\neg\Box\neg\varphi \mid : \mid : \neg\neg\varphi} \, (\neg\Box L_2)}{\neg\Box\neg\varphi : \mid : \neg\Box\neg\varphi} \, (\neg\Box R)}{\neg\Box\neg\varphi : \Box\neg\Box\neg\varphi} \, (\Box R)}{: \neg\Box\neg\varphi \to \Box\neg\Box\neg\varphi} \, (\to R)$$

7.6.3 Derivability of connexive $S5$-non-theorems

I now show the impact of the modified falsification condition of '$\Box$', by deriving in $SCS5$ some connexive non-theorems of $S5$.

First, the propositional connexive axioms are derivable. Below is a derivation for A_1; the other propositional connexive axioms have similar derivations.

$$\cfrac{\cfrac{\cfrac{\varphi : \varphi}{\varphi : \neg\neg\varphi} \, (\neg\neg R)}{: \neg(\varphi \to \neg\varphi)} \, (\neg\to R)}{}$$

Proposition 7.6.33.

(1) $\vdash_{SCS5} \neg(\Box\varphi \to \Box\neg\varphi)$ (2) $\vdash_{SCS5} \neg(\neg\Box\varphi \to \Box\varphi)$

Proof. (1)

$$\cfrac{\cfrac{\cfrac{: \mid \neg\varphi : \neg\varphi}{\neg\Diamond\varphi : \mid : \neg\varphi} \, (\neg\Diamond L)}{\neg\Diamond\varphi : \Diamond\neg\varphi} \, (\Diamond R_1)}{: \neg\Diamond\varphi \to \Diamond\neg\varphi} \, (\to R)$$

(2) Similar and omitted. □

Proposition 7.6.34.
$$\vdash_{SCS5} \neg\Box\varphi \equiv \Box\neg\varphi \qquad (7.6.19)$$

Proof. I show that $\vdash_{SCS5} \neg\Box\varphi \to \Box\neg\varphi$. The other direction is similar and omitted.

$$\cfrac{\cfrac{\cfrac{: \ | \ \neg\varphi : \neg\varphi}{\neg\Box\varphi : \ | \ : \neg\varphi} \ (\neg\Box L_2)}{\neg\Box\varphi : \Box\neg\varphi} \ (\Box R)}{: \neg\Box\varphi \to \Box\neg\varphi} \ (\to R)$$

□

7.6.4 About $SC5$ models

I am not presenting a full model-theory for $CS5$, as I am not concerned here with completeness proofs. However, it is quite clear how to go about such a model-theory.

A frame would consist only of a set points P, without any accessibility relation. The support for truth and falsity for '$\Box$' in a model $\mathcal{M}$ and point p would be the following.

$$\mathcal{M}, p \vDash^+ \Box\varphi \text{ iff } \forall q \in P: \mathcal{M}, q \vDash^+ \varphi$$

$$\mathcal{M}, p \vDash^- \Box\varphi \text{ iff } \forall q \in P: \mathcal{M}, q \vDash^- \varphi$$

7.6.5 Linguistic support for $CS5$

In this section, I present some linguistic motivating support for the falsification conditions of the connexive modalities employed by $CS5$.

7.6 The connexive modal logic $CS5$

The support is relies again on a distinctions of meaning based on *intonational stress* (in speech) in a dialogue. Recall the other connections of intonational stress with connexivity, for the poly-connexive logic $PCON$, as presented in Section and Francez [37].

First, note that in natural language there are different possible patterns of locating the intonational stress in a sentence expressing a modal proposition. I will indicate the intonationally stressed part by surrounding it with *focal brackets*: $\langle...\rangle_F$; those brackets do not belong to the object-language[4] and are used for comprehensibility only. Thus, contrast the following intonational stress patterns:

$$(neutral\ stress)\ \langle \Box \varphi \rangle_F$$
$$(focal\ content)\ \Box \langle \varphi \rangle_F$$
$$(focal\ modality)\ \langle \Box \rangle_F \varphi$$

The different intonational stress patterns are exemplified by the following sentences:

(1) ⟨It will necessarily rain tomorrow⟩$_F$

(2) It will necessarily ⟨rain tomorrow⟩$_F$

(3) It will ⟨necessarily⟩$_F$ rain tomorrow

Consider, in turn, each of the above sentences asserted by a participant in a dialogue.

1. The neutral stress[5] expresses a "plain" modal assertion. There is no common background of the dialogue participants involved.

2. The focal content pattern expresses a situation where there is certainty (i.e., it is part of the common ground of the dialogue participants) about the *necessity of something*, but there are alternatives as to what might be necessary.

---
[4] Though it may be interesting to incorporate them.
[5] This might as well be represented as the absence of any intonational stress.

In (2), there is a certainty about the necessity regarding the weather the next day. The assertion adds what is the assumed necessary content, in this case of it raining the next day. The alternative content is it not raining the next day.

3. The focal modality expresses a situation where there is certainty about *some content*, but there are alternatives as to the modal status of this content.

   In (3), there is certainty about it raining the next day. The assertion adds that this content is necessary. The alternative is that it raining the next day is (only) possible.

The major observation is, that *a difference in intonational stress leads to a difference in negating the modality*.

1. Neutral stress leads to classical $S5$ modal negation, also having a neutral stress.
$$\langle \neg \Box \varphi \rangle_F \equiv \langle \Diamond \neg \varphi \rangle_F$$

2. Focal stress leads to the negating of necessity as proposed for $CS5$, namely (7.5.7), preserving the focal stress.
$$\neg \Box \langle \varphi \rangle_F \equiv \Box \langle \neg \varphi \rangle_F \qquad (7.6.20)$$

Example 7.6.35. *Alice and Bob are economists, both sharing the view that the recent economic crisis will necessarily have an impact on the interest rate, but they disagree about what that effect is.*

> Alice : The interest rate will necessarily $\langle$rise$\rangle_F$.
> Bob : No! The interest rate will necessarily $\langle$decline$\rangle_F$
> $\qquad\qquad\qquad\qquad\qquad\qquad\qquad\qquad$ (7.6.21)

where 'decline' *is* not rise.

By uttering 'No', Bob negates Alice's assertion, and continues by explicitly stating what this negation amounts to. It amounts to

7.6 The connexive modal logic $CS5$

negating the (intonationally stressed) contents, replacing it by the alternative, its negation.

□

More interestingly, the impact of connexive modal negation is even more strongly recognizable[6] in case the negated modal assertion is the consequent of a connexive conditional.

Example 7.6.36. *Alice and Bob are both fans of the football team T. They both share the belief that the expected rain next day will necessarily affect the result of tomorrows's match, but disagree as to what the effect will be.*

$Alice$: if it rains tomorrow, team T will necessarily $\langle\text{win}\rangle_F$.
Bob : No! if it rains tomorrow, team T will necessarily $\langle\text{loose}\rangle_F$
(7.6.22)
where 'loose' is not win).

By uttering 'No', Bob negates Alice's assertion (the whole conditional), and continues by explicitly stating what this negation amounts to. It amounts to another conditional, which, being connexive, has the negated original modal consequent as its consequent.

That is,

$$\neg(\varphi \to \Box\langle\psi\rangle_F) \equiv^{A_1} \varphi \to \neg\Box\langle\psi\rangle_F \equiv^{(7.6.20)} \varphi \to \Box\langle\neg\psi\rangle_F$$

□

3. Focal modality leads to yet another way of negating the modal, related to pragmatic implicature.

$$\neg\langle\Box\rangle_F \varphi \equiv \langle\Diamond\rangle_F \varphi \qquad (7.6.23)$$

Example 7.6.37.

$Alice$: It will $\langle\text{necessarily}\rangle_F$ rain tomorrow.
Bob : No! It will (only...) $\langle\text{possibly}\rangle_F$ rain tomorrow

---
[6]I thank Heinrich Wansing for alerting me to this point.

This negation could lead to another version of a connexive $S5$, by taking *it* as the falsification condition. I will not pursue this possibility further here.

7.6.6 Possibility

I now turn to the consideration of the possibility operator '$\Diamond$'. The most natural way to incorporate possibility into $CS5$ is under the focal contents approach.

Example 7.6.38. *Once again, Alice and Bob are economists, this time both sharing the view that the recent economic crisis will possibly have an impact on the interest rate, but they again disagree about that effect.*

$Alice$: The interest rate will possibly $\langle \text{rise} \rangle_F$.
Bob : No! The interest rate will possibly $\langle \text{decline} \rangle_F$ (i.e., not rise)
$$\tag{7.6.24}$$
As in Example 7.6.35, by uttering 'No', Bob negates Alice's assertion, and continues by explicitly stating what this negation amounts to. It amounts to negating the (intonationally stressed) content, replacing it by the alternative, its negation.

□

This leads to the modality rules for '$\Diamond$' in Figure 7.2.

Thus, negating possibility is according to the following scheme, the analog of (7.5.7).

Proposition 7.6.35.
$$\vdash_{SCS5} \neg\Diamond\varphi \equiv \Diamond\neg\varphi \tag{7.6.29}$$

Proof. I show that $\vdash_{SCS5} \neg\Diamond\varphi \to \Diamond\neg\varphi$. The other direction is similar and

7.6 The connexive modal logic $CS5$

$$\frac{\mathcal{G} \mid \Gamma, \varphi : \Delta \mid \varphi :}{\mathcal{G} \mid \Gamma, \Diamond \varphi : \Delta} \ (\Diamond L) \tag{7.6.25}$$

$$\frac{\mathcal{G} \mid \Gamma : \Delta \mid \Sigma : \varphi, \Theta}{\mathcal{G} \mid \Gamma : \Delta, \Diamond \varphi \mid \Sigma : \Theta} \ (\Diamond R_1) \qquad \frac{\mathcal{G} \mid \Gamma : \Delta, \varphi}{\mathcal{G} \mid \Gamma : \Delta, \Diamond \varphi} \ (\Diamond R_2) \tag{7.6.26}$$

$$\frac{\mathcal{G} \mid \Gamma, \neg \varphi : \Delta \mid \neg \varphi :}{\mathcal{G} \mid \Gamma, \neg \Diamond \varphi : \Delta} \ (\neg \Diamond L) \tag{7.6.27}$$

$$\frac{\mathcal{G} \mid \Gamma : \Delta \mid \Sigma : \neg \varphi, \Theta}{\mathcal{G} \mid \Gamma : \Delta, \neg \Diamond \varphi \mid \Sigma : \Theta} \ (\neg \Diamond R_1) \qquad \frac{\mathcal{G} \mid \Gamma : \Delta, \neg \varphi}{\mathcal{G} \mid \Gamma : \Delta, \neg \Diamond \varphi} \ (\neg \Diamond R_2) \tag{7.6.28}$$

Figure 7.2: Modal rules for '$\Diamond$' in $SCS5$

omitted.

$$\frac{\dfrac{: \mid \neg \varphi : \neg \varphi}{: \neg \Diamond \varphi \mid \neg \varphi :} \ (\neg \Diamond R_1)}{\dfrac{\neg \Diamond \varphi : \Diamond \neg \varphi}{: \neg \Diamond \varphi \to \Diamond \neg \varphi} \ (\to R)} \ (\Diamond L)$$

$\square$

Proposition 7.6.36.

(1) $\vdash_{SCS5} \ : \neg(\Diamond \varphi \to \Diamond \neg \varphi)$ (2) $\vdash_{SCS5} \ : \neg(\Diamond \neg \varphi \to \Diamond \varphi)$

Proof. Again, I only show a derivation for (1).

$$\frac{\dfrac{\varphi : \mid \varphi : \varphi}{\dfrac{\varphi : \mid \varphi : \neg \neg \varphi}{\dfrac{\varphi : \neg \Diamond \neg \varphi \mid \varphi :}{\dfrac{\Diamond \varphi : \neg \Diamond \neg \varphi}{: \neg (\Diamond \varphi \to \Diamond \neg \varphi)} \ (\neg \to R)} \ (\Diamond L)} \ (\neg \Diamond R_2)} \ (\neg \neg R)}$$

$\square$

Note that the usual duality between necessity and possibility *does not* hold in $CS5$. The loss of duality in connexive modal logic is not new. See, for example, $CS4$, a connexive $S4$-based modal logic in Kamide and Wansing [51].

Finally, there is also a possibility of negating '$\Diamond$' according to the focal modality, dually to (7.6.23).

$$\neg \langle \Diamond \rangle_F \varphi \equiv \langle \Box \rangle_F \varphi \qquad (7.6.30)$$

As stated above, I will not pursue this option further here.

Chapter 8

Conclusion

I have presented a more or less bird's eye view of Connexive logics, an important family of contra-classical logics.

8.1 Uncovered topics

While having presented many of the central topics related to connexivity, some important topics were not covered, though. For example:

- Decidability of validity in the various connexive logics. For example, Wansing's C, MC and CK are decidable, shown by an embedding into their respective positive fragments, which are known to be decidable. See Wansing [122] for a discussion.

- Ternary accessibility relations based model theory for Connexive logics.

    Such ternary accessibility elations were introduced by Routley and Meyer [104] for defining models of Relevance logics.

Frames based on ternary accessibility relations were employed for providing a model-theory for Connexive logics by Routley [103] and Mortensen [75].

In [124], Wansing and Unterhuber use the Chellas-Segerberg semantics, a relation between two states and a set of states. This may be seen as generalizing a frame semantics with a ternary relation between three states.

- The effect of the presence of '$\bot$' and '$\top$' in the object-language of a Connexive logic.

  One such study is Pizzi [92], who consider instances of the connexive axioms with '$\bot$' and '$\top$'.

- Counterfactuals.

  The proof-theory and model-theory involve features that I could not describe in the intended space for the book. The interested reader may consult, for example, Pizzi [91].

- Probability and probabilistic logics.

  There are interesting connections between a connexive conditional and probability theory, in particular with the notion of conditional probability. I consider those topics beyond he intended scope of the book. The reader might consult Pfeifer [90].

8.2 Topics for further research

There are issues related to connexivity that are open to further research. I list below several of them.

8.2.1 Connexivity and content

As already emphasized above (p. 12), it is commonly agreed that connexivity is based on some *connection in content* between the antecedent and consequent of a valid conditional. In spite of this consensus, there is by now *no explicit proposal of a theory of content and of the nature of the connection*, a theory that *leads* to a Connexive logic.

For example, due to the lack of such a suitable theory, Pizzi [91, p. 284] writes:

> I will make the assumption however that a logic of connection may be formulated without a previous analysis of the notion of connection; on the contrary, it is plausible to think that a logic of connection may throw light on the notion of connection itself.

For example, there is no definite view about the question whether the connection in content between the antecedent and the consequent should be transitive. The axiom

$$\vdash (\varphi \to \psi) \to ((\psi \to \chi) \to (\varphi \to \chi))$$

included in many Relevance logics and Connexive logics, can be interpreted as a imposing such transitivity.

Similarly to the situation with Relevance logics there are attempts to reduce connection in content to the syntactic criterion of the variable-sharing condition (VSC), for example the work of Weiss [126].

Here is what Pizzi [92, p.127] has to say about (VSC) as the connection in content, an opinion I share too:

> The idea of requiring that the antecedent and the consequent should have something in common in order to grant

an interconnection between them is a sound principle, which however classical relevantists have forced into the rigid syntactical criterion of variable-sharing.

My own opinion is, that a suitable notion of content and its sharing should be obtained by abstraction from the notion of *meaning* in natural language. By now, there are in the literature two competing major theories of meaning:

- *Model-theoretic semantics*, based on *truth* (in a model) as its central notion, which is the prevailing theory.
- *Proof-theoretic semantics*, based on *proof* (or derivation) in a meaning-conferring proof-system, a more revisionary theory; see Francez [32] for a detailed presentation.

There is, though, a newly emerging theory of meaning, with *information* (and its flow) as the central concept. The informational view provides a natural interpretation to the notions of 'support of truth' and 'support of falsity', used in many model-theoretical accounts of connexivity.

Most relevant in our context is Shramko and Wansing's [112] (in particular, Section 2), that provides an illuminating relationship between connection in content and information.

Further research along those lines seems promising.

8.2.2 Connexivity and bilateralism

The centrality of the role of negating the conditional in obtaining connexivity calls for *bilateralism*, a theory of meaning based both on *ver-*

8.2 Topics for further research

ification and *falsification* as primitive operations[1], as well as viewing truth and falsity as primitive notions, not each defined by absence of the other. In a bilateral formal proof system there are two kinds of formal derivations: *proofs* and *disproofs* (called also *dual proofs*). Proofs correspond to validity, while disproofs correspond to *dual validity* (specified below).

A first and important step in the direction of a bilateral approach to connexivity is by Wansing [121], defining the logic $2C$, a *bi-connexive* logic, a connexive variant of his bi-intuitionistic logic $2Int$. The logic is defined via a natural-deduction system containing two kinds of rules: rules for proofs and rules for disproofs, leading to the above mentioned two kinds of derivations.

The logic is referred to as bi-connexive because in addition to the connexive implication it has also another connexive conditional, the connexive *co-implication* '$\twoheadleftarrow$', a dual of the connexive implication. The formula $\psi \twoheadleftarrow \varphi$ is read as 'φ co-implies ψ'. There doesn't seem to be any NL counterpart for co-implication.

Models for $2C$ are similar to those of C. The following clauses for support of truth and support of falsity are added:

$$\mathcal{M}, t \models^+ \psi \twoheadleftarrow \varphi \text{ iff } \forall s \geq t: \text{ either } \mathcal{M}, s \not\models^- \varphi \text{ or } \mathcal{M}, s \models^+ \psi$$

$$\mathcal{M}, t \models^- \psi \twoheadleftarrow \varphi \text{ iff } \forall s \geq t: \text{ either } \mathcal{M}, s \not\models^- \varphi \text{ or } \mathcal{M}, s \models^- \psi$$

Dual validity of φ in a model $\mathcal{M}$ of $2C$, denoted $\mathcal{M} \models^d \varphi$, is defined by

$$\mathcal{M} \models^d \varphi \text{ iff for every } t: \mathcal{M}, t \models^- \varphi$$

Dual validity of φ in $2C$, denoted $\models^d_{2C} \varphi$, holds if φ is dually valid in every model $\mathcal{M}$ of $2C$.

---

[1]More often, the basis of bilateralism is regarded as constituting of two primitive *speech acts*, *assertion* and *denial*.

The co-implication dually validates the duals of the connexive axioms.

$$co - A_1 : \vDash^d_{2C} \neg(\neg\varphi \twoheadleftarrow \varphi)$$

$$co - A_2 : \vDash^d_{2C} \neg(\varphi \twoheadleftarrow \neg\varphi)$$

$$co - B_1 : \vDash^d_{2C} (\neg\psi \twoheadleftarrow \varphi) \twoheadleftarrow \neg(\psi \twoheadleftarrow \varphi)$$

$$co - B_2 : \vDash^d_{2C} \neg(\psi \twoheadleftarrow \varphi) \twoheadleftarrow (\neg\psi \twoheadleftarrow \varphi)$$

Like C, $2C$ is negation inconsistent, but non-trivial. For any φ, both $\vdash_{2C} \varphi \wedge \neg\varphi \to \neg(\varphi \wedge \neg\varphi)$ and $\vdash_{2C} \neg(\varphi \wedge \neg\varphi \to \neg(\varphi \wedge \neg\varphi))$ both hold.

8.2.3 Provable contradictions

As mentioned several times, Connexive logics tend to be negation inconsistent. However, mostly those negation inconsistent logics are paraconsistent, so the presence of contradiction does not lead to triviality.

We have encountered several such provable contradictions.

- In C:
  $$(i) \vdash_C \varphi \wedge \neg\varphi \to \varphi \quad (ii) \vdash_C \neg(\varphi \wedge \neg\varphi \to \varphi)$$
- In $2C$ (and also in C):
  $\vdash_{2C} \varphi \wedge \neg\varphi \to \neg(\varphi \wedge \neg\varphi)$ and $\vdash_{2C} \neg(\varphi \wedge \neg\varphi \to \neg(\varphi \wedge \neg\varphi))$
- In $\mathcal{L}_{rc}$:
  $$\vdash_{\mathcal{N}_{rc}} (((\neg\varphi \to \varphi) \to \neg\varphi) \to \neg((\neg\varphi \to \varphi) \to \neg\varphi))$$

contradicting A_1.

However, *not every contradiction is provable* in a negation inconsistent Connexive logics. Certainly $\nvdash_C p \wedge \neg p$ and $\nvdash_{\mathcal{N}_{rc}} p \wedge \neg p$ for an atomic formula p.

8.2 Topics for further research

So, a natural question arising is the following: For any given negation inconsistent connective logic, say $\mathcal{L}$, *which contradictions are provable in $\mathcal{L}$?*

Characterizing provable contradictions may lead to a better understanding of connexivity.

Chapter 9

Glossary: linguistics terms

This glossary contains brief explanations and exemplifications of some common terms from linguistics, brought for self-containment and for the benefit of readers not acquainted with those terms.

abbreviations: • VP: verb phrase.

constituent: In the syntactic analysis of NL sentences, there are legitimate groupings and illegitimate ones. The grouping is represented by nested bracketing, where brackets are indexed by syntactic categories. For example, in the sentence 'John loves Mary', the usual bracketing is

$$[_S \text{John} \, [_{VP} \text{loves Mary}]]$$

rendering loves Mary as a constituent of the sentence (of syntactic category VP). On the other hand, the grouping

$$(*) \, [_S [_? \text{John loves}] \text{Mary}]]$$

is an incorrect grouping (indicated by '*'), hence John loves is not a constituent of the above sentence.

coordination particles: Those are the NL counterparts of connectives in logic, words like and or, that are regimented sometimes as conjunction and disjunction in logics. However, the repertoire of NL coordination particles is much richer than the standard stock , including, for example, *contrastive coordination particles* like but and although.

A major difference between NL coordination particles and logical connectives is that the latter apply to closed formulas (the regimentation of full sentences), while the former apply also to *sub-sentential phrases* (parts of sentences), e.g.:

- *noun phrases*, like John and Marry, or every girl and some boy.
- *verb phrases*, like sing or dance, eat bred and drink water.

In particular, coordination can take place also between non-constituents, as shown in the third item of (4.3.25).

intonational stress: This is a phenomenon of *spoken* language, represented by $[_F...]$ when written. The phrase braced has an intonational stress expressing *focus*. Semantically, focus is usually associated with alternatives. For example, in

$$[_F \text{John}] \text{ loves Mary}$$

John is in focus, implying that it is John, and, say, not Bill or Ed that loves Mary. On the other hand, in a different focalization such as

$$\text{John loves } [_F \text{Mary}]$$

Mary is in focus, implying that John loves Mary and not, say, Sue or Sara.

conjunction reduction: A process of *distributing* a coordination particle over a quantifier, like in (4.3.26) and (4.3.27), which sometimes preserves logical equivalence and sometimes does not.

expansion: This is not a standard term. It also means distribution of a coordination particle, raising it to sentential coordination. For example, an expansion of

> Mary sings and dances

yields
> Mary sings and Mary dances

infelicity: In contrast to logic, where that has only two levels of specification, namely syntax and semantics, NL have more levels. For examples, pragmatics. One such level is that of *felicity*. An expression may pass the syntactic and semantic filters, but still be deficient in some other way, related to as *infelicity*. Most often, infelicity is related to some *triviality*, rendering a sentence non-assertable. One of the best examples of infelicity is Hurford's constraint on disjunction. Consider

> Mary lives in France or she lives in Paris

This sentence is clearly syntactically well-formed, and semantically interpretable as a disjunction. However, since the second disjunct implies the first, it is considered infelicitous.

Bibliography

[1] Alan R. Anderson and Nuel D. Belnap Jr. *Entailment (vol. 1)*. Princeton University Press, N.J., 1975.

[2] Richard B. Angell. A propositional logic with subjunctive conditionals. *Journal of symbolic logic*, 27:327–343, 1962.

[3] Richard B. Angell. Tre logiche dei condizionali congiuntivi. In Claudio Pizzi, editor, *Leggi di natura, modalit, ipotesi. La logica del ragionamento controfattuale*. Milan: Feltrinelli, 1978. Translation by the author of Three logics of subjunctive conditionals, paper presented August 12, 1966 at Colloquium on Logic and Foundations of Mathematics, Hannover, Germany.

[4] Arnon Avron. Simple consequence relations. *Information and Computation*, 92:105–139, 1991.

[5] Arnon Avron. Whither relevance logic. *Journal of philosophical Logic*, 21(3):243–281, 1992.

[6] Jc Beall. There is no logical negation: true, false, both and neither. In Patrick Girard and Zach Weber, editors, *Non-classicality: logic, philosophy, mathematics*. 2016, forthcoming. Special issue of the Australasian Journal of Logic.

[7] Jc Beall, Ross T. Brady, Allen P. Hazen, Graham Priest, and Greg Restall. Relevant restricted quantification. *Journal of Philo-*

sophical Logic, 35(6):587–598, 2006. DOI:10.1007/s10992-005-9008-5.

[8] Nuel D. Belnap. How a computer should think. In Gilbert Ryle, editor, *Contemporary aspects of philosophy*, pages 30–56. Stocksfield:Oriel Press, 1976.

[9] Nuel D. Belnap. A useful four-valued logic. In J. Michael Dunn and George Epstein, editors, *Modern uses of multiple-valued logic*, pages 8–37. Dordrecht:Reidl, 1977.

[10] Jean-Yves Béziau. Paraconsistent logic and contradictory viewpoint. *Revista Brasileira de Filosofia*, 241, 2014.

[11] Jean-Yves Béziau, Walter A. Carnielli, and Dov M. Gabbay, editors. *Handbook of Paraconsistency*. College Publications, London, 2007.

[12] James R. Bode. The possibility of a conditional logic. *Notre Dame Journal of Formal Logic*, XX(1), 1979.

[13] John Cantwell. The logic of conditional negation. *Notre Dame Journal of Formal Logic*, 49(3):245–260, 2008.

[14] William S. Cooper. The propositional logic of ordinary discourse 1. *Inquiry*, 11(1-4), 1968.

[15] Vincenzo Crupi and Andrea Iacona. The evidential conditional. *Erkenntnis*, 2020. DOI https://doi.org/10.1007/s10670-020-00332-2.

[16] Kosta Došen. The first axiomatization of relevant logic. *Journal of Philosophical Logic*, 21(4):339–356, 1992.

[17] P. B. Downing. Subjunctive conditionals, time order, and causation. *Proceedings of the Aristotelian Society, n.s.*, 59, 1958-1959.

[18] P. B. Downing. Opposite conditionals and deontic logic. *Mind*, 70(280):491–502, 1961.

[19] Michael Dummett. *The Logical Basis of Metaphysics*. Harvard University Press, Cambridge, MA., 1993 (paperback). Hard copy 1991.

[20] J. Michael Dunn. Intuitive semantics for first-degree entailments and coupled trees. *Philosophical Studies*, 29:149–168, 1976.

[21] J. Michael Dunn and Greg Restall. Relevance logic. In Dov M. Gabbay and Franz Guenther, editors, *Handbook of philosophical logic*, pages 1–136. Kluwer, 2002. Vol. 6 - 2nd edition.

[22] Paul Egré and Guy Politzer. On the negation of indicative conditionals. In Maria Aloni, Michael Franke, and Floris Roelofsen, editors, *Proceedings of the 19th Amsterdam Colloquium*, pages 10–18, 2013.

[23] Luis Estrada-González. Weakened semantics and the traditional square of opposition. *Logica Universalis*, 2(1), 2008.

[24] Luis Estrada-González. The Bochum plan and the foundations of contra-classical logics. *CLE e-Prints*, 19(1), 2020.

[25] Luis Estrada-González. Possibility, consistency, connexivity. In Nicola Olivetti, Rineke Verbrugge, Sara Negri, and Gabriel Sandu, editors, *13th Conference on Advances in Modal Logic, AiML 2020, Helsinki, Finland*, pages 189–207. College Publications, London, UK, 2020.

[26] Luis Estrada-González and Elisángela Ramrez-Cámara. A comparison of connexive logics. *IfCoLog Journal of Logics and their Applications*, 2016. (Special issue on Connexive Logic, to appear).

[27] Thomas Macaulay Ferguson. A computational interpretation of conceptivism. *Journal of Applied Non-Classical Logics*, 24(4):333–367, 2014. DOI10.1080/11663081.2014.980116.

[28] Thomas Macaulay Ferguson. Ramsey's footnote and priest's connexive logics. *Bulletin of Symbolic Logic*, 20:387–388, 2014. An abstract of a paper presented at ASL Logic Symposium 2012.

[29] Thomas Macaulay Ferguson. Logics of nonsense and parry systems. *Journal of Philosophical Logic*, 44(1):65–80, 2015. DOI: 10.1007/s10992-014-9321-y.

[30] Thomas Macaulay Ferguson. On arithmetic formulated connexively. *IFCoLog Journal of Logics and Their Applications*, 3(3):357–3876, 2016.

[31] Thomas Macaulay Ferguson. *Meaning and proscription in formal logic*. Springer, Cham, Switzerland, 2017.

[32] Nissim Francez. *Proof-theoretic Semantics*. College Publications, London, 2015.

[33] Nissim Francez. Natural-deduction for two connexive logics. *IfCoLog Journal of Logics and their Application*, 3(3):479–504, 2016. Special issue on Connexive Logic.

[34] Nissim Francez. A proof-theoretic semantics for adjectival modification. *Journal of Logic, Language and Information*, 26(1), 2017. 10.1007/s10849-016-9245-8.

[35] Nissim Francez. Connexive restricted quantification. *Notre Dame Journal of Formal Logic*, 61(3), 2020. DOI 10.1215/00294527-2020-0015.

[36] Nissim Francez. A poly-connexive logic. *Logic and Logical Philosophy*, 29(1), 2020. DOI: 10.12775/LLP.2019.022.

[37] Nissim Francez. Poly-connexivity: connexive conjunction and disjunction. *Notre Dame Journal of Formal Logic*, 2020. under refereeing.

[38] Dov M. Gabbay. What is negation in a system 2020? In Ofer Arieli and Anna zamansky, editors, *Arnon Avron on Semantics and Proof Theory of Non-Classical Logics, Outstanding Contributions to Logic series*. Springer, 2021.

[39] Jovana Gacić. Negated definite conjunction and its implicatures. In M. Teresa Espinal, Elena Castroviejo, Louise McNally, and Cristina Real-Pugdillers, editors, *Proceedings of Sinn und Bedeutung 23, vol. 1*. Universitat Autònoma de Barcelona, 2019.

[40] Guido Gherardi and Eugenio Orlandelli. Super-strict implications. *Bulletin of the Section of Logic*, 2021. http://dx.doi.org/10.18778/0138-0680.2021.2.

[41] Siegfried Gottwald. Many-valued logic. In Edward N. Zalta, editor, *The Stanford Encyclopedia of Philosophy*. Metaphysics Research Lab, Stanford University, summer 2020 edition, 2020.

[42] H. Paul Grice. Logic and conversation. In Peter Cole and Jerry. L. Morgan, editors, *Syntax and Semantics 3*. Academic Press, New York, NY, 1975.

[43] H. Paul Grice. *Studies in the way of words*. Harvard University Press, Cambridge, Mass., 1989.

[44] Chung-Hye Han and Maribel Romero. Disjunction, focus, and scope. *Linguistic Inquiry*, 35(2):179–217, 2004.

[45] Simon T. Hewitt. Need anything follow from a contradiction? *Inquiry*, 2020. https://doi.org/10.1080/0020174X.2020.1762728, to appear.

[46] Lawrence R. Horn and Heinrich Wansing. Negation. In Edward N. Zalta, editor, *The Stanford Encyclopedia of Philosophy*. Summer 2015 edition edition, 2015.

[47] James Hurford. Exclusive or inclusive disjunction. *Foundations of Language*, 11:409–411, 1974.

[48] Andrea Iacona. Strictness and connexivity. *Inquiry (forthcoming)*, 2019. DOI https://doi.org/10.1080/0020174X.2019.1680428.

[49] Thomas Jarmużek and Jacek Malinowski. Boolean connexive logics: Semantics and tableau approach. *Logic and Logical Philosophy*, 28(3):427–448, 2019. DOI: 10.12775/LLP.2019.003.

[50] Tomasz Jarmużek and Bartosz J. Kaczkowski. On some logic with a relation imposed on formulae: tableau system F. *Bulletin of the Section of Logic*, 43(1/2):53–72, 2014.

[51] Norihiro Kamide and Heinrich Wansing. Connexive modal logic based on positive S4. In Jean-Yves Béziau and Marcelo Conigli, editors, *Logic without Frontiers*, pages 389–409. College Publications, King's College, London, 2011. Festschrift for Walter Alexandre Carnielli on the Occasion of His 60th Birthday.

[52] Norihiro Kamide and Heinrich Wansing. *Proof Theory of N4-related Paraconsistent Logics*. College Publications, London, 2015. Studies in Logic, Vol. 54.

[53] Norihiro Kamide and Heinrich Wansing. Completeness of connexive heyting-brouwer logic. *FCoLog Journal of Logic and their Applications*, 3(3):441–486, 2016.

[54] Andreas Kapsner. Strong connexivity. *thought*, 1:141–145, 2012.

[55] Andreas Kapsner. *Logics and Falsification*. Springer, 2014.

[56] Andreas Kapsner. Humble connexivity. *Logic and logical philosophy*, 28:513–536, 2019. Special Issue: Advances in Connexive Logic. Guest Editors: Hitoshi Omori and Heinrich Wansing, DOI: http://dx.doi.org/10.12775/LLP.2019.001.

[57] Andreas Kapsner. Connexivity and the pragmatics of conditionals. *Erkenntnis*, 2020. DOI https://doi.org/10.1007/s10670-020-00325-11.

[58] Andreas Kapsner and Hitoshi Omori. Will the real Boethius please stand up? *The Reasoner*, 14(3):13–14, 2020.

[59] William Kneale and Martha Kneale. *The Development of Logic*. Duckworth, London, 1962.

[60] Wolfgang Lenzen. A critical examination of the historical origins of connexive logic. *History and Philosophy of logic*, 28, 2019. doi.org/10.1080/01445340.2019.16506101.

[61] Clarence I. Lewis and Cooper H. Langford. *Symbolic Logic*. Dover, New York, 1932.

[62] Paul Lorentzen. *Einfürung in die Logic und Mathematik*. Springer, Berlin, Germany, 1955. Second edition, 1969.

[63] Paul Lorentzen and Kuno Lorenz. *Dialogische Logik*. Wissenschaftliche Buchgesellschaft, Darmstadt, Germany, 1978.

[64] Jean Łuasiewicz. *Aristotles Syllogistic from the Standpoint of Modern Formal Logic*. Oxford University Press, Oxford,, UK, 1951.

[65] Oana Lungu, Anamaria Fălăus, and Francesca Panzeri. Disjunction in negative contexts: a cross-linguistic experimental study. *Submitted*, 2019.

[66] Jacek Malinowski and Rafał Palczewski. Relating semantics for connexive logic. In Alessandro Giordani and Jacek Malinowski, editors, *Logic in High Definition*, pages 49–65. Springer, Cham, 2020.

[67] Edwin Mares. Negation. In Leon Horsten and Richard Pettigrew, editors, *The continuum companion to philosophical logic*, pages 180–215. Continuum international publishing group, London, New York, 2011.

[68] Edwin Mares and Francesco Paoli. C.I. Lewis, E.J. Nelson, and the modern origins of connexive logic. *Organon F*, 26(3):405–426, 1966. https://doi.org/10.31577/orgf.2019.26304.

[69] Edwin D. Mares. Relevance and conjunction. *Journal of Logic and Computation*, 22:7–21, 2012.

[70] Christopher J. Martin. Embarrassing arguments and surprising conclusions in the development of theories of the conditional in the twelfth century. In Jean Jolivet and Alain De Libera, editors, *Gilbert De Poitiers et ses Contemporains*, pages 377–401. Naples, 1987.

[71] Storrs McCall. Connexive implication. *Journal of symbolic logic*, 31:415–433, 1966.

[72] Storrs McCall. Connexive class logic. *Journal of symbolic logic*, 32(4):83–90, 1967.

[73] Storrs McCall. Connexive Gentzen. *Logic Journal of IGPL*, 22(6):964–981, 2012. doi: 10.1093/jigpal/jzu019.

[74] Storrs McCall. A history of connexivity. In Dov M Gabbay, Francis J. Pelletier, and John Woods, editors, *Handbook of the history of logic, Volume 11: Logic: a history of its central concepts*, pages 415–449. Elsevier, Amsterdam, 2012.

[75] Chris Mortensen. Aristotle's thesis in consistent and inconsistent logics. *Studia Logica*, 43:107–116, 1984. https://doi.org/10.1007/BF00935744.

[76] Sara Negri and Jan von Plato. Sequent calculus in natural deduction style. *Journal of Symbolic Logic*, 66(4):1803–1816, 2001.

[77] David Nelson. Constructible falsity. *ournal of Symbolic Logic*, 14(11):16–26, 1949.

[78] David Nelson. Constructible falsity. *Journal of Symbolicl Logic*, 14:16–26, 1949.

[79] Everett J. Nelson. Intensional relations. *Mind*, 39(156):440–453, 1930.

[80] Ricardo Arturo Nicolás-Francisco. A note on three approaches to connexivity. *Archives of Philosophy, Say*, (51):129–137, 2019. DOI: 10.26650/arcp2019-5108.

[81] Sergey P. Odintsov, Daniel Skurt, and Heinrich Wansing. Connexive variants of modal logics over FDE. In Ofer Arieli and Anna Zamansky, editors, *Arnon Avron on Non-Classical Logics - Between Semantics and Proof Theory*. Springer, 2021, To appear.

[82] Grigory K. Olkhovikov and Peter Schroeder-Heister. On flattening general elimination rules. *Review of symbolic logic*, 7(1), 2014.

[83] Hitoshi Omori. From paraconsistent logic to dialetheic logic. In Holger Andreas and Paul Verdèe, editors, *Logical Studies of Paraconsistent Reasoning in Science and Mathematics*, pages 111–134. Springer, 2016.

[84] Hitoshi Omori. A note on Francez' half-connexive formula. *IFCoLog Journal of Logic and their Applications*, 3(3):505–512, 2016. Special issue on Connexive Logic.

[85] Hitoshi Omori. Towards a bridge over two approaches in connexive logic. *Logic and logical philosophy*, 2019. DOI: 10.12775/LLP.2019.005.

[86] Hitoshi Omori and Heinrich Wansing. 40 years of FDE: An introductory overview. *Studia Logica*, 105:1021–1049, 2017. Special issue.

[87] Hitoshi Omori and Heinrich Wansing. On contra-classical variants of Nelson logic N4 and its classical extension. *The Review of Symbolic Logic*, 11(4):805–820, 2018.

[88] Hitoshi Omori and Heinrich Wansing. An extension of connexive logic c. In Nicola Olliveti, Rineke Verbrugge, Sara Negri, and Gabriel Sandu, editors, *Advances in Modal Logic, Vol 13 (AiML)*, pages 503–522. College Publications, London, 2020.

[89] Niki Pfeifer. Experiments on aristotle's thesis: Towards an experimental philosophy of conditionals. *The Monist*, 95(2):223–240, 2012.

[90] Niki Pfeifer. Probability logic. In Markus Knauff and Wolfgang Spohn, editors, *Handbook of Rationality*. MIT press, Cambridge MASS., 2021, To appear.

[91] Claudio Pizzi. Boethius thesis and conditional logic. *Journal of Philosophical Logic*, 6(1):283–240, 1977.

[92] Claudio Pizzi. Aristotle's thesis between paraconsistency and modalization. *Journal of Applied Logic*, 3(1):119–131, 2005.

[93] Claudio Pizzi and Timothy Williamson. Strong boethius thesis and consequential implication. *Journal of Philosophical Logic*, 26(5):569–588, 1997.

[94] Jan Von Plato. Gentzen's proof systems: byproducts of the work of a genius. *The Bulletin of Symbolic Logic*, 18(3):313–367, 2012.

[95] Francesca Poggiolesi. A cut-free simple sequent calculus for modal logic S5. *Review of Symbolic Logic*, 1(1):3–15, 2008.

[96] Graham Priest. Negation as cancellation, and connexive logic. *Topoi*, 18(2):141–148, 1999.

[97] Graham Priest. *An Introduction to Non-Classical Logic (2nd edition)*. Cambridge University Press, 2008.

[98] Graham Priest. *An Introduction to Non-Classical Logics: From If to Is (2nd edition)*. Cambridge University Press, 2008.

[99] Shahid Rahman and Helge Rückert. Dialogical connexive logic. *Synthese*, 127:105–139, 2001.

[100] Frank P. Ramsey. General propositions and causality. In David Hugh Mellor, editor, *Philosophical Papers*. Cambridge University Press, 1990.

[101] Greg Restall. Relevant and substructural logics. In Dov Gabbay and John Woods, editors, *Handbook of the history of logic: Volume 7, Logic and modalities in the twentieth century*. Elsevier, 2006.

[102] Mats Rooth. A theory of focus interpretation. *Natural Language Semantics*, 1(1):75–116, 1992. https://doi.org/10.1007/BF02342617.

[103] Richard Routley. Semantics for connexive logics. *Studia Logica*, 37(4):393412, 1978.

[104] Richard Routley and Robert K. Meyer. he semantics of entailment (i). In Hugues Leblanc, editor, *Truth, Syntax, and Modality*, pages 199–173. North-Holland, Amsterdam, 2002.

[105] Richard Routley and Hugh A. Montgomery. On systems containing aristotle's thesis. *The Journal of Symbolic Logic*, 33(1):82–96, 1968.

[106] Richard Routley and Valerie Routley. Negation and contradiction. *Revista Columbiana de Mathemáticas*, 19(1-2):201–230, 1985.

[107] Gillian Russel. Logical pluralism. In Edward N. Zalta, editor, *The Stanford Encyclopedia of Philosophy*. Metaphysics Research Lab, Stanford University, summer 2019 edition, 2019.

[108] Peter Schroeder-Heister. A natural extension of natural deduction. *Journal of symbolic logic*, 49:1284–1300, 1984.

[109] Peter Schroeder-Heister. The categorical and the hypothetical: a critique of some fundamental assumptions of standard semantics. *Synthese*, 187(3):925–942, 2012.

[110] Peter Schroeder-Heister. The calculus of higher-level rules, propositional quantification, and the foundational approach to proof-theoretic harmony. In Andrzej Indrzejczak, editor, *Gentzen's and Jaśkowski's Heritage: 80 Years of Natural Deduction and Sequent Calculi*. 2014. Special issue of Studia Logica. DOI: 10.1007/s11225-014-9562-3.

[111] Sebastian Sequoiah-Grayson. Dynamic negation and negative information. *Review of Symbolic Logic*, 2(1):233–248, 2009.

[112] Yaroslav Shramko and Heinrich Wansing. The nature of entailment: an informational approach. *Synthese*, 2019. https://doi.org/10.1007/s11229-019-02474-5, corrected in Synthese 2000.

[113] Peter F. Strawson. *Introduction to logical theory*. London: Methuen, 1952.

[114] Richard Sylvan. A preliminary western history of sociative logics. In *Bystanders Guide to Sociative Logics, Research Series in Logic and Metaphysics # 9 (Chapter 4)*. Australian National University, Canberra, 1989. Published as chapter 5 of *Sociative Logics and their Applications*. Essays by the Late Richard Sylvan, D. Hyde and G. Priest (eds.), Aldershot: Ashgate Publishing, 2000.

[115] Anna Szabolcsi and Bill Haddican. Conjunction meets negation: a study in crosslinguistic variation. *J. of Semantics*, 21(3):219–249, 2004.

[116] Bruce E. R. Thompson. Why is conjunctive simplification invalid? *Notre Dame Journal of formal logic*, 32(2):248254, 1991.

[117] Mathieu Vidal. When conditional logic met connexive logic. In Clair Gardent and Christian Retoré, editors, *Proceedings of*

the 12th International Conference on Computational Semantics (IWCS), 2017.

[118] Pilar Terrés Villalonga. From natural to formal language: A case for logical pluralism. *Topoi*, 38(2):333–345, 2017. DOI 10.1007/s11245-017-9490-8.

[119] Heinrich Wansing. Connexive modal logic. In Renate Schmidt, Ian Pratt-Hartmann, Mark Reynolds, and Heinrich Wansing, editors, *Advances in modal logic*, volume 5, pages 367–383. College Publications, King's College, London, 2005.

[120] Heinrich Wansing. Natural deduction for bi-connexive logic and a two-sorted typed λ-calculus. *IFCoLog Journal of Logics and their Applications*, 3(3):413–439, 2016.

[121] Heinrich Wansing. On split negation, strong negation, information, falsification, and verification. In Katalin Bimbo, editor, *J. Michael Dunn on Information Based Logics*, pages 161–189. Springer, 2016. volume 8 of Outstanding Contributions to Logic.

[122] Heinrich Wansing. Connexive logic. In Edward N. Zalta, editor, *The Stanford Encyclopedia of Philosophy*. Fall 2020 edition, 2020. https://plato.stanford.edu/archives/spr2020/entries/logic-connexive/.

[123] Heinrich Wansing and Daniel Skurt. Negation as cancellation, connexive logic, and qLPm. *Australasian Journal of Logic*, 13(2):476488, 2018.

[124] Heinrich Wansing and Matthias Unterhuber. Connexive conditional logic. part I. *Logic and Logical Philosophy*, 28(3):5674610, 2019. DOI: 10.12775/LLP.2018.018.

[125] Yale Weiss. Connexive extensions of regular conditional logic. *Logic and Logical Philosophy*, 28(3):611–627, 2019. DOI: http://dx.doi.org/10.12775/LLP.2018.012.

[126] Yale Weiss. Semantics for pure theories of connexive implication. *Review of Symbolic Logic*, 2020. Forthcoming.

[127] J.E Wiredu. A remark on a certain consequence of connexive logic for zermelos set theory. *Studia Logica*, 33:127130, 1974.

Names Index

Łukasiewicz, Jean, 40

Abelard, Peter, 23, 53
Anderson, Alan Ross, 13, 56, 129–131, 133, 135, 137
Angell, Richard B., 15, 30, 44, 65, 68, 69, 71, 97, 104
Aristotle, 20, 27, 28, 40
Avron, Arnon, 131, 132, 141, 144, 147

Béziau, Jean-Yves, 12, 84
Beall, JC, 150, 166
Belnap, Nuel D., 13, 38, 56, 129–131, 133, 135, 137, 182
Bode, James R., 41
Boethius, Anicius Manlius Severinus, 21, 23
Brady, Ross T., 150

Cantor, Georg, 155
Cantwell, John, 28, 29, 59, 72, 76, 77, 79

Carnielli, Walter A., 12
Caroll, Lewis, 101
Chellas, Brian F., 98, 198
Cooper, William S., 4, 29, 72, 77, 78, 92, 95
Crupi, Vincezo, 104

De Morgan, Augustus, 106, 110
Došen, Kosta, 136
Downing, P. B., 31
Dummett, Michael, 82
Dunn, Michael J., 38, 144, 145, 182

Egré, Paul, 179
Estrada-González, Luis, 19, 26, 44, 98, 177, 179, 180
Euclid, 40

Fălăuș, Anamaria, 124
Ferguson, Thomas M., 21, 37, 40, 60, 69, 80, 149, 167
Francez, Nissim, 32, 72, 96, 106, 111, 127, 130, 134, 149, 185, 191, 200

Gabbay, Dov M., 12
Gać, Jovana, 122
Gentzen, Gerhard, 133
Gherardi, Guido, 62
Gottwald, Siegfried, 15
Grice, Paul H., 37

Haddican, Bill, 121
Han, Chang-Hye, 124
Hazen, Allen P., 150
Hewitt, Simon T., 11
Hilbert, David, 108, 143
Horn, Lawrence R., 82
Hurford, James, 3

Iacona, Andrea, 104, 179

Jarmużek, Tomasz, 100

Kaczkowski, Bartosz J., 100
Kamide, Norihiro, 31, 32, 84, 109, 119, 196
Kapsner, Andreas, 1, 5, 23, 26, 38, 104, 105, 178, 179
Kleene, Stephen C., 38
Kneale, William and Martha, 21
Kripke, Saul, 15, 50, 186

Langford, Cooper H., 179
Lenzen, Wolfgang, 1
Lewis, Clarence I., 41, 179
Lorentzen, Paul, 99
Lorenz, Kuno, 99
Lungu, Oana, 124

Mackinson, David, 105

Malinowski, Jacek, 100, 101
Mares, Edwin, 41, 97, 140, 147
Martin, Christopher J., 1, 39
McCall, Storrs, 1, 21, 23, 40, 41, 45, 97, 104, 106, 137, 155, 157, 161, 169, 174
Meyer, Robert K., 197
Montgomery, Hugh A., 40, 69
Mortensen, Chris, 56, 198

Negri, Sara, 84
Nelson, David, 9, 38, 48, 109, 119
Nelson, Everett J., 36, 41, 97
Nicolás-Francisco, Ricardo Arturo, 180

Odintsov, Sergey, 15, 181, 184, 185
Olkhovikov, Grigory K, 112
Omori, Hitoshi, 23, 45, 59, 90, 98, 109, 119, 127, 182
Orlandelli, Eugenio, 62
Orlov, Ivan E., 136

Palczewski, Rafał, 101
Panzeri, Francesca, 124
Paoli, Francesco, 41, 97
Pfeifer, Niki, 42, 198
Pizzi, Claudio, 23, 28, 31, 97, 198, 199
Poggiolesi, Francesca, 186, 187
Politzer, Guy, 179
Post, Emil, 10
Prawitz, Dag, 132

NAMES INDEX

Priest, Graham, 10, 12, 38, 60, 63, 98, 99, 150, 152, 157, 167, 168, 174, 181, 183

Rückert, Helge, 99, 106
Rahman, Shahid, 99, 106
Ramírez-Cámara, Elisángela, 19, 26, 98
Ramsey, Frank P., 76, 80
Restall, Greg, 136, 144, 145, 150
Romero, Maribel, 124
Rooth, Mats, 119
Routley, Richard, 10, 22, 40, 69, 129, 197, 198
Routley, Valerie, 10
Russell, Bertrand, 153
Russell, Gillian, 7

Schroeder-Heister, Peter, 112, 131
Segerberg, Krister, 198
Sequoiah-Grayson, Sebastian, 73
Shramko, Yaroslav, 200
Skurt, Daniel, 15, 60, 181, 184, 185
Strawson, Peter F., 30

Sylvan, Richard see Routley, Richard, 22
Szabolcsi, Anna, 121

Terrés Villalonga, Pilar, 37
Thompson, Bruce E. R., 36

Unterhuber, Matthias, 22, 98, 198

Vidal, Mathieu, 105
von Plato, Jan, 84, 133

Wajsberg, Mordechai, 138
Wansing, Heinrich, 1, 15, 20, 22, 27, 31, 32, 36, 41, 43–45, 48, 50, 58–60, 77, 82, 84, 98, 99, 107, 109, 119, 127, 129, 130, 143, 152, 157, 158, 160, 174, 176, 178, 181, 182, 184, 185, 193, 196–198, 200, 201
Weiss, Yale, 23, 130, 131, 136, 199
Williamson, Timothy, 23, 31, 97
Wiredu, J. E., 150

Logics Index

$2C$ (Wansing), 60, 201, 202
$2Int$ (Wansing), 201
B, 56
BK^- (Odintsov, Skurt and Wansing), 181
C (Wansing), 36, 45–48, 50, 51, 53, 55–59, 62, 77, 130, 143, 152, 158, 160, 161, 185, 197, 201, 202
$C3$ (Omori and Wansing), 59
CC (Wansing), 158
$CC1$ (McCall), 21, 71, 97
CK (Wansing), 181, 197
CN (Cantwell), 28, 59, 72
$CS4$ (Kamide and Wansing), 196
$CS5$ (Francez), 185, 187, 190, 192, 194
FDE, 38, 59, 98, 181–183
$KFDE$ (Priest), 181, 183, 184
K_3 (Kleene), 38
LP (Priest), 38, 98
MC (Wansing), 58, 197

MRS^P (Estrada-González), 98
$N3$ (Nelson), 38
$N4$ (Nelson), 38, 119
NL (Nelson), 97
OL - Ordinary language logic (Cooper), 29, 72, 92, 95
PA_1 (Angell), 65, 67, 68, 71, 97
PA_2 (Angell), 71
$PCON$ (Francez), 106–109, 114, 118–120, 126, 128, 191
P_N (Priest), 167
P_S (Priest), 63, 64, 152, 157, 167, 174
P_S^p (Priest, Francez), 171, 172
QC (Wansing), 46, 152, 157–161, 166, 167, 174, 175
QP_S (Priest), 152, 167–171, 175
QP_S^p (Priest, Francez), 152, 175
$S4$, 196

$S5$, 186, 187, 189
$\mathcal{L}_{rc}$ Connexive Relevant logic (Francez), 130, 131, 135, 136, 141, 143–145, 147
cBK^- (Odintsov, Skurt and Wansing), 184
cCL (Wansing and Unterhuber, 98
R (Anderson and Belnap), 130, 131, 135, 138–140, 144, 147, 148
dLP (Omori), 98

Bi-Connexive logic (Wansing), 176
BiInt–Bi-Intuitionistic logic, 31

Classical logic, 1, 6–12, 15, 20, 21, 28, 30, 31, 35, 36, 38, 46, 59, 62, 69, 83, 100, 105, 107, 119, 149, 150, 152, 154, 157, 161, 166, 169
Conditional logic(s), 23
  Connexive conditional logics, 98
Connexive logic(s), 1, 5, 7, 9, 10, 12, 14–16, 19, 20, 27, 28, 32–35, 38–40, 43–45, 59, 60, 65, 72, 77, 86, 89, 95–98, 103, 104, 120, 129, 130, 136, 149, 150, 152, 157, 158, 180, 181, 197–199, 202

Bi-connexive logic(s), 201
Boolean Connexive logic, 100, 101
Connexive modal logic(s), 15, 180, 181, 184
Dialogical Connexive logic, 99, 106
First-order Connexive logic(s), 27, 150, 157, 158, 167, 174, 175
Hyper-Connexive logic(s), 22, 29, 31, 76, 94
Kapsner-strong Connexive logic(s), 26
Material Connexive logic MC (Wansing), 58
Poly-Connexive logic, 106, 107
Constructive logic(s), 9
Containment logic(s), 37
Contra-Classical logic(s), 8, 9, 21, 33, 39, 44, 197

Intuitionistic logic, 9, 10, 38, 46, 59, 75, 108, 144

Modal logic(s), 14, 15, 177, 180, 186
  K, 181
Multi-valued logics, 15

N4 (Nelson), 9, 48, 109
Non-Classical logic(s), 7, 9

OL - Ordinary language logic (Cooper), 77

LOGICS INDEX

Paracomplete logic(s), 10, 12, 60, 151, 160, 172, 175
Paraconsistent logic(s), 10, 12, 36, 58, 60, 151, 160, 172, 175, 183, 185, 202
Philosophical logic(s), 1
Probabilistic logic(s), 198

Relevance logic(s), 12–14, 56, 129–132, 135, 144, 147, 150, 197, 199

Sub-Classical logic(s), 8, 9, 13
Supra-Classical logic(s), 8, 14

Subject Index

Absurdity ('⊥'), 75
Accessibility relation, 15, 98, 186, 190
    Ternary accessibility relations, 197

Biconditional, 24, 31, 62, 144
    Connexive biconditional, 46, 49
Bilateralism, 200, 201
Binary quantification (see Restricted quantification), 149
Bochum plan, 44, 184
Boolean algebra, 155

Closure
    Closure of derivations under composition, 84
Co-implication, 31, 60, 201
    Connexive co-implication, 60, 201
Compatibility, 97
Conditional, 1, 4, 5, 12–14, 16, 20, 24, 28–34, 37–42, 44, 47, 58, 60, 63, 64, 72–78, 80–84, 91, 92, 95, 97–99, 103–107, 128–132, 134, 141, 142, 150, 179, 184, 187, 199, 200
    Asymmetry of, 24
    Connexive conditional, 23, 35, 46–48, 50, 58, 62, 65, 100, 103, 104, 127, 149, 155, 175, 184, 193
    Connexive-relevant Conditional, 140, 141
    Consequential conditional, 97
    Counterfactual conditional, 23
    Embeddable conditional, 181
    Indicative conditional, 28, 30, 41, 42
    Intensional conditional, 97
    Material conditional, 1, 12, 13, 23, 28, 38, 42, 46,

65, 76, 130, 149, 155
  Paradoxes of the material
    conditional, 13, 65, 69
  Nested conditional, 76
  Relevant conditional, 14,
    133, 147
  Strict conditional, 179
  Subjunctive conditional,
    31, 65
  Super-strict conditional, 62
  Transitivity of the
    conditional, 34
conditional, 2
Conditional excluded middle,
  98
Conjunction, 4, 10, 11, 23, 34,
  35, 107, 111, 118–120,
  122, 144, 147, 150
  Conjunction reduction, 120
  Conjunction simplification,
    34–36, 40, 41, 70, 97
  Connexive conjunction,
    106, 114
  Multiplicative conjunction,
    36
Connexive algebra, 155, 157,
  161, 163, 164,
  168–170, 172, 174
Connexive class theory, 149,
  150
Consistency/Inconsistency, 10,
  41, 55, 65, 105, 136
  Negation consistency, 65,
    71
  Negation inconsistency, 10,

12, 36, 37, 58, 87, 90,
  99, 137, 202
Post inconsistency, 10
Self-consistency, 179
Content, 12, 13, 30, 60, 74, 92,
  97, 107, 128–130, 141,
  199
Contradiction, 9–12, 14, 21, 36,
  39, 60, 73, 75, 81, 83,
  91, 105, 151, 202
  Provable contradiction, 202
  Self-contradiction, 30
Contradictory, 80, 104, 105
  Self-contradictory, 105
Contraposition, 34, 38, 57, 62,
  94, 95
Coordination, 120
  Constituent coordination,
    120, 126, 127
  Non-constituent
    coordination, 120
  Sentential coordination,
    121, 127
  Sub-sentential
    coordination, 120
Coordination stress, 126
Counterfactuals, 1, 198

Decidability, 197
Deduction-theorem, 141, 142
  Natural deduction-theorem,
    142
    Relevant-connexive
      natural-deduction-
      theorem,

142
Derivability/Non-derivability, 16, 22, 24, 72, 75–77, 86, 89, 91, 94, 96, 109, 135
Dialetheism, 12
Discharge index, 131
Disjunction, 4, 9, 11, 20, 35, 36, 71, 107, 111, 118–120, 124–126, 147
    Connexive disjunction, 106, 114
    Disjunction addition, 35–37, 71
    The disjunction property, 9
Disjunctive syllogism, 40
Double negation, 9, 26, 38, 62, 75, 78, 79, 83

Ex Contradictione Nihil (ECN), 11
Ex Contradictione Quodlibet (ECQ), 10
Ex falso quodlibet (EFQ), 14
Ex impossibily quodlibet, 24
Excluded middle, 9
Explosion, 10, 12, 71, 95, 105, 106, 140

Falsification, 201
Falsification condition(s), 44, 45, 47, 48, 50, 58, 60, 63, 99, 107, 111, 119, 127, 134, 159, 185–187, 189, 190

Felicity conditions, 2
Fission, 147
Focus, 118, 119
Formal languages, 4
Fusion, 147, 148

Gap, 15, 29, 166, 182
Glut, 15, 166, 182
Gricean maxims, 37

Humble connexivity, 104, 178
Hyper-sequent, 186, 187
Hyper-sequent calculus, 186

Identity, 25, 67
Implication (see Conditional), 16
Implicature, 2, 3
Implosion, 11
Information, 50, 200
Intensional, 62, 97, 161
Intonational stress, 3, 4, 79, 118–120, 122–128, 185, 191, 192
Involutive, 83

Logical consequence, 5, 8, 10–12, 14, 20, 23, 30, 36, 38, 45, 50, 58, 60, 62–64, 69, 105, 110, 141, 151, 160, 167, 171, 172, 175, 183, 184
    Non-monotonic logical consequence, 63
Logical monism, 7
Logical pluralism, 7, 151

Logical truth, 6
Logical validities, 8

Modality, 4
Models, 15
    Kripke models, 15
Modus ponens, 25, 34, 47, 108, 134, 145, 156
Modus tollens, 111
Multiple conclusions, 11

Natural language (NL), 1–6, 8, 12, 17, 28, 30, 37, 41, 65, 72, 74, 76, 80, 82, 92, 96, 107, 118, 120, 126, 128, 134, 186, 200, 201
Natural-deduction (ND), 14, 31, 35, 37, 45, 72, 75, 81, 91, 96, 130, 131, 135, 142, 147, 201
Necessarium ex quodlibet, 25
Necessary, 14
Necessity, 177, 179, 180, 185, 196
Neg raising, 82
Negation, 2, 5, 16, 20, 21, 24, 29, 31, 34, 38, 42, 60, 62, 72–77, 81–83, 91, 92, 95, 96, 103, 104, 107, 111, 125, 128, 131
    Constructive negation, 9
    Corrective negation, 74
    Demi-negation, 119
    Negation as cancellation, 60

Negation concord, 119
Non-uniform negation, 82
Strong negation, 10, 38, 46, 48, 58, 62, 166, 175

Paracompleteness, 12, 55, 56, 60, 62–64, 157
Paraconsistency, 12, 55, 56, 60, 62–64, 84, 157
Possibilism, 180
Possibility, 177, 179, 180, 185, 194, 196
Possible, 14
Post-completeness, 8
Pragmatics, 2, 5, 37
Presupposition, 2, 76
Principles of subjunctive contrariety, 65
Proof-theoretic semantics, 96
Proposition, 5, 12–14, 16, 20, 50, 73, 96
    Atomic proposition, 16, 73, 76, 78, 81, 82, 91, 92
Proscriptive principle, 37

Quantifier(s), 150

Ramsey's test, 80
    Ramsey's dual test, 80
Reductio ad absurdum, 9, 21, 23
Relatedness semantics, 100
Relevance/irrelevance, 36, 37
Restricted quantification, 149, 150, 154, 160

Semantics

Model-theoretic semantics, 200
Proof-theoretic semantics, 200
Sequent, 186, 187
Square of opposition, 98
Strong connexivity, 26, 54
Structural rule(s), 14
    Contraction, 187
    Cut, 187
    Weakening, 14, 187
Sub-contrary, 73, 76, 82, 84

Tautology, 6, 11
Truth-condition(s), 64
Truth-functionality, 13, 35, 38, 74, 91, 97
Truth-table(s), 59, 65, 68, 71, 91, 92, 97, 180, 182
Truth-value(s), 12, 15, 29, 41, 61, 68, 76, 77, 82, 83, 91, 94, 129, 181, 182
    Designated truth-values, 68, 91, 94

Uniform substitution, 25, 82, 92, 100, 109

Vacuity, 61
Variable-sharing condition (VSC), 13, 131, 199
Verification, 201
Verification condition(s), 45, 48, 60, 63, 159

www.ingramcontent.com/pod-product-compliance
Lightning Source LLC
Chambersburg PA
CBHW070732160426
43192CB00009B/1410